Won for All

**How the *Drosophila* Genome
Was Sequenced**

OTHER TITLES FROM COLD SPRING HARBOR
LABORATORY PRESS

Ageless Quest: One Scientist's Search for the Genes That Prolong Youth
The Eighth Day of Creation: The Makers of the Revolution in Biology
Mendel's Legacy: The Origin of Classical Genetics
Inspiring Science: Jim Watson and the Age of DNA
The Inside Story: DNA to RNA to Protein
Drosophila: *A Laboratory Handbook,* 2nd Edition
Fly Pushing: The Theory and Practice of Drosophila *Genetics*

Won for All

How the *Drosophila* Genome Was Sequenced

Michael Ashburner

Portraits by Lewis Miller

Epilogue by R. Scott Hawley
Stowers Institute for Medical Research

Afterword by Ethan Bier
University of California, San Diego

 COLD SPRING HARBOR LABORATORY PRESS
Cold Spring Harbor, New York

Won for All
How the *Drosophila* Genome Was Sequenced

Publisher	John Inglis
Acquisition Editors	John Inglis and Alex Gann
Development Director	Jan Argentine
Project Coordinator	Mary Cozza
Permissions Coordinator	Maria Fairchild
Production Editor	Rena Steuer
Desktop Editor	Susan Schaefer
Production Manager	Denise Weiss
Cover Designer	Lewis Agrell

Cover artwork: *Drosophila* image reprinted, with permission, from Morgan T.I I. and Bridges C.B. 1916. *Sex-linked inheritance* in Drosophila. Carnegie Institution of Washington, Publication No. 237, Washington, D.C. Panel shows the content of the fly genome as represented in chromosome arm 2R. Each predicted gene is depicted as a box color-coded by similarity with genes from the mouse, *C. elegans,* and *S. cerevisiae.*

Library of Congress Cataloging-in-Publication Data

Ashburner, M.
  Won for all : how the Drosophila genome was sequenced / Michael Ashburner; epilogue by R. Scott Hawley ; afterword by Ethan Bier.
     p. cm.
  ISBN 0-87969-802-0 (hardcover : alk. paper)
  1. Drosophila--Genome mapping--History. I. Title.

  QL537.D76A844 2006
  572.8'633--dc22
                            2005032174

10 9 8 7 6 5 4 3 2

All Cold Spring Harbor Laboratory Press publications may be ordered directly from Cold Spring Harbor Laboratory Press, 500 Sunnyside Blvd., Woodbury, New York 11797-2924. Phone: 1-800-843-4388 in continental U.S. and Canada. All other locations: (516) 422-4100. FAX: (516) 422-4097. E-mail: cshpress@cshl.edu. For a complete catalog of all Cold Spring Harbor Laboratory Press publications, visit our World Wide Web site http://www.cshlpress.com/.

For David Hogness,
who taught us so much

Contents

The Portraits

The portraits of some of those who played a part in this story were drawn by Lewis Miller, the Australian artist. Lewis was born in 1959 in Melbourne and studied painting at the Victorian College of the Arts. His works are owned by the National Gallery of Australia in Canberra, the National Gallery of Victoria, and by many other public and private collectors. He has traveled extensively and in 2003 was Australia's Official War Artist in Iraq. Lewis was Artist-in-Residence at Cold Spring Harbor Laboratory in 1998 and 2000, where his portrait of James Watson hangs in the Grace Auditorium. In 1998, he was awarded Australia's highest honor for portrait painters, the Archibald Prize.

Most of the drawings in this book were executed during the Human Genome Symposium at Cold Spring Harbor Laboratory in May 2003. Others were drawn at Cold Spring Harbor during the Genome Informatics meeting in October 2005. John Sulston was drawn at the Sanger Institute, Hinxton, in June 2003.

Preface

One of the wonders of modern science is our ability to determine the complete sequence of genomes of organisms as different, qualitatively and in size and complexity, as a small bacterial pathogen and a human. The technical breakthrough was made in the mid-1970s by Fred Sanger at Cambridge and Wally Gilbert at Harvard. It is, however, only in about the last 15 years that high-volume automated sequencing has been possible, largely the result of pioneering work by Lee Hood, then at Caltech. The family of automated sequencing machines that resulted from this work, ABI machines, was developed and sold by Perkin-Elmer (PE; now the Applied Biosystems division of Applera). These developments, and others, have revolutionized biology. As I write this preface in the summer of 2005, the complete genomic sequences of nearly 300 bacteria have been determined, as have those of at least 50 more complex eukaryotes. The ultimate prize—the draft sequence of the human genome— was published in February 2001, simultaneously by a publicly funded international consortium in *Nature* and by a company, Celera, of which more is written below, in *Science*.

The sequencing of the human genome was very political.[1] It may make the story that follows seem like a nursery tale, but

---

[1] See Sulston J. and Ferry G. 2002. *The common thread: A story of science, politics and ethics and the human genome.* Bantam Press, London; Shreeve J. 2004. *The genome war: How Craig Venter tried to capture the code of life and save the world* Alfred Knopf, New York.

nonetheless interesting for that. This story is concerned with events that transpired during a 2-year period, from May 1998 to March 2000, that led to the "complete" sequence of the genome of the small fly, *Drosophila melanogaster*, an organism devoid of any direct economic or medical relevance, but one of a very small number of "model" organisms whose study has, during the last 90 years, had a major impact on our understanding of genetics and basic biological processes. Despite its size, more than 5000 people worldwide make their living by studying this fly, more than 150 in Cambridge alone (whenever I write Cambridge without qualification, I mean the original one, in the Fens). It had been obvious for some years that we had to determine this important organism's genomic sequence, and two groups, one in Europe and one in the U.S., began to do this in 1995. Both groups were following the classic strategy of sequencing individual clones that had been first physically mapped to the genome. Given a few years, and a vital injection of cash,[2] we would have finished the job.

In May of 1998, we were surprised, to say the least, by an announcement by Dr. J. Craig Venter that a newly formed company, Celera Genomics, intended to sequence the genome of this fly. They planned to do so using a method called random whole-genome "shotgun" sequencing, untried on this scale. This announcement led to a formal collaboration between the publicly funded U.S. *Drosophila* sequencing group and this new company, with the Europeans tagging along. Relationships between academia, with its long and solid tradition of the public release of information at the earliest opportunity, and commercial companies, with responsibilities to their shareholders, are often fraught. The aims of the two sides may appear to coincide, but ideas about the means of attain-

---

[2] Although the U.S. effort was well funded by the National Institutes of Health (NIH), we had to engage the truly Byzantine DGXII of the European Commission, which, to quote Sydney Brenner in another context, might be said to "have vision but poor eyesight."

ing those aims, and the compromises that must be made on the way, differ.

I was fortunate (or unfortunate) to be close to, though not at, the epicenter of this collaboration. There are a number of ways in which this story could have been written. I have chosen to do so in a sub-Chandleresque style, because I found that this suited my purpose, which, frankly, was therapy. The reader will understand that although I permit myself a degree of poetic license, the important facts are as close to reality as I can make them. I have relied not only on my memory, and on the memories of one or two trusted colleagues, but also on my notes from meetings, papers, diaries, and e-mails.

The "let it all hang out" school of scientific reporting was, of course, pioneered by Jim Watson in *The Double Helix* (1968). The discovery of the structure of DNA was enormously more significant than the events recounted here. Nevertheless, the absence of a scholarly apparatus in the original edition of this book (partly corrected in the Norton Critical Edition [edited by Gunther S. Stent, W.W. Norton, New York, 1980]) made many of the stories rather impenetrable to an outsider. Partly for this reason, partly because it was pretentious, and partly because I thought that the footnotes were the best part of my school Chaucer, I have attempted to provide this story with reasonably explicit footnotes.

There is no sex in this story. There are several reasons for this, but I will state only one. My housemaster taught me that certain matters are best not discussed in public. Because I have no idea who, if anyone, will read this, it seemed best to leave certain matters unsaid. There is, however, and I apologize to those readers who are sensitive to such matters, a certain amount of what, in polite society, is often considered to be bad language. But to censor that would be to be unfaithful to both my sources and emotions.

The story is, of course, incomplete. Chapter 1 was written in one weekend and the subsequent chapters in a couple of additional sessions. I have no intention of following this story up; the catharsis is complete. My thanks, however, go to those who did read it and to those mentioned in the pages that follow who gave me the opportunity to be there, at least some of the time. To those nearest me who have suffered, here is why.

A NOTE ON "NOW" AND "THEN"

The great bulk of this work was written over a few frantic weekends in early 2001, but it was only in April 2005 that I showed it to Jim Watson and John Inglis at Cold Spring Harbor Laboratory with the idea of publication. I wrote this account in the time of the action portrayed, and the positions people are said to hold are those that they held at that time. If, in a note, I use the word "now," I am referring to August 2005, when I revised this work for publication. I have done this rather sparingly, but in the Postscript, I have given brief accounts of the present whereabouts and doings of most of those mentioned.

Acknowledgments

Several friends read a draft of this book and encouraged me to publish it. Thanks to Suzanna Lewis, Martin Reese, Gerry Rubin, and Mark Yandell for their help and for correcting many errors. All errors that remain are, of course, my own. I thank John Inglis, Alex Gann, and Jim Watson at Cold Spring Harbor Laboratory for both encouragement and more practical help. I am also grateful to Jim for making Lewis Miller's drawings available. Thanks to Wally Gilbert, Peter Lawrence, Kitsos Louis, Jan Witkowski, and the staff of the Carnegie Library at CSHL for information and support. I am grateful to Ken Burtis and Garry Karpen, who provided photographs of the jamboree.

At Cold Spring Harbor Laboratory Press it has been a great pleasure to work again with Jan Argentine, Editorial Development Director, Denise Weiss, Production Manager; Susan Schaefer, Desktop Editor; Mary Cozza, Project Coordinator; and Rena Steuer, Production Editor.

1

Opening Gambits

"Have *you* heard the news?"

Have I heard the news? What does he think? I have spent the last 12 hours traveling, 90 minutes listening to stories about golf on the Isle of Man from Ken driving from Cambridge to Heathrow, 7 hours stuck in seat 42H next to an obese man with piles on British Airways to Kennedy, and 45 minutes dozing on the fake leather seats of MarWood's[3] Lincoln stuck in rush hour traffic on the optimistically named Long Island Expressway. "Have I heard the *news*?"

"No, but you are going to tell me."

"Craig[4] is going to sequence human by shotgun privately and sell it and do *Drosophila* in 10 days as a warm up." Breathless.

Turmoil? Everyone is rushing around like headless chickens. Jim is calling foul, Frank is is apoplectic, and Eric[5] is shouting at

---

[3] Kenway Cars of Sawston, Cambridge drives European Bioinformatics Institute (EBI) staff to and from the airports. Ken is the boss, and he spends his summers playing golf on the Isle of Man. If Ken drives you, never ask him about golf. Try chickens. If John drives you, ask him to sing his version of "The Teddy Bear's Picnic." I have it on good authority (Suzanna Lewis; see footnote 48) that it is quite fine and passes the time stuck in the traffic jam on the M25. MarWood is a limo service that provides the only civilized option for getting from J.F.K. Airport to Cold Spring Harbor Laboratory. But you need an introduction; they don't drive just anybody.

[4] J. Craig Venter, then head of Celera Genomics. Either the hero or the villain of this story. Described as a former "California beach bum" by Steve Connor in *The Independent*, December 1, 1999.

[5] James D. Watson, of double helix fame. Chancellor of Cold Spring Harbor Laboratory, previously its president. Was the first head of the U.S. Human

---

Michael Ashburner

some poor graduate student that she should stop wasting time at Cold Spring Harbor,[6] hotfoot it back to MIT, and get some real sequencing done. Michael and John[7] are said to be arriving from Hinxton on the Concorde.[8] Wade[9] is rushing around getting quotes from anyone, simply anyone. Gerry? Gerry's[10] imme-

---

Genome Program (National Center for Human Genome Research). Jim had little time for Craig, for it was over the issues of cDNA sequencing (championed by Craig) and the patenting of these sequences (championed by Bernadine Healy, then director of the National Institutes of Health) that Jim left the NIH. He jumped, rather than having been pushed, but the latter was only a matter of time (see J.B. Wyngaarden, in Inglis J.R., Sambrook J., and Witkowski J.A., eds. 2003. *Inspiring science: Jim Watson and the age of DNA*, pp. 377–385. Cold Spring Harbor Laboratory Press, Cold Spring Harbor, New York). Were I his obituarist, I would write "Jim never suffered fools gladly." Francis Collins is a human geneticist and was Jim's successor as director of the National Institute for Human Genome Research. Francis is tall, rides a red Honda Nighthawk 750, and entertains his audiences with his own compositions, accompanying himself on the guitar. Eric Lander is a man who got bored with mathematics and teaching business, and had Peter Cherbas (see footnote 121) teach him biology. Either Peter taught him well, he was a good student, or both. Then head of the genome lab at the Whitehead Institute at MIT.

[6] Cold Spring Harbor Laboratory, on the north shore of Long Island, is one of the world's premier locations for modern biological research. Famous for its courses and conferences. You either love it (I do) or hate it (few do). Good bird-watching in the harbor, and a resident osprey (*Pandion haliaetus*) nests each year on the spit; you occasionally see black skimmers (*Rynchops niger*) in the inner harbor.

[7] Michael Morgan was then chief executive of the Wellcome Trust Genome Campus at Hinxton, ten miles south of Cambridge, England. John Sulston was then director of the Sanger Centre (now Sanger Institute) at Hinxton. The Sanger Centre sequenced one third of the human genome. Both Michael and John were strong and powerful proponents of the idea that sequence data should be freely in the public domain.

[8] They did not; rather, they arrived, but not on the Concorde.

[9] Nick Wade, a science correspondent for the *New York Times*. Gave Craig endless free publicity.

[10] Gerald M. Rubin, then John D. MacArthur Professor of Genetics at the University of California, Berkeley; during the jamboree (see below), he was announced as the new vice president for biomedical research at the Howard Hughes Medical Institute. The leader of the Berkeley *Drosophila*

Craig Venter

diate comment is too politically incorrect to quote here. It was surprising, for a Berkeley professor. In essence, he made it clear that he was placed in the position of a reluctant groom whose future father-in-law marches him up the aisle with a loaded shotgun in his hand. The idea of collaborating with Craig was not immediately appealing, but the alternative was much worse.

"Can he do it?" "No way," says Phil,[11] inventor of PHRED and PHRAP.[12] "Maybe," say more cautious souls. "Look at his track record: ESTs,[13] *Haemophilus*,[14] building TIGR[15] into a major sequencing lab. Maybe he just can."

Genome Project, which, together with the much smaller European *Drosophila* Genome Project (lead by David Glover, then my boss in the department of genetics at Cambridge), was sequencing the fly genome, in Gerry's case funded by the NIH via the National Human Genome Research Institute (NHGRI). A hero of this story.

[11] Phil Green invented the programs used by most users of ABI377 machines to check sequence quality and assemble the raw reads into finished sequence. Phil is a distinguished computational biologist at the University of Washington.

[12] Two eponymously named pieces of software (these small shows of vanity are quite commonplace in science) that determine the As, Cs, Gs, and Ts of a sequence read (PHRED) and then match these small strings of letters up where they are the same in order to patch them into larger strings (PHRAP).

[13] Expressed sequence tags are short hits of sequence from messenger RNAs, the intermediates between the DNA and protein sequences. Craig championed EST sequencing at NIH. It was a good idea, but strongly opposed by Jim Watson on the grounds that it would distract attention (and funds) from the larger project, which was to sequence the genome itself.

[14] *Haemophilus influenzae*, a bacterium, does not cause influenza (that is the result of a virus). See Fleischmann R.D., Adams M.D., White O., Clayton R.A., Kirkness E.F., Kerlavage A.R., et al. 1995. Whole-genome random sequencing and assembly of *Haemophilus influenzae* Rd. *Science* 269: 496–512.

[15] The Institute of Genome Research (TIGR), a not-for-profit outfit founded by Craig when he left the NIH after they refused to fund his human EST project. TIGR was first financed by Human Genome Sciences, a commercial company headed by Bill Haseltine (who was not Jim's graduate student, despite what you may hear). Craig and Haseltine had a dramatic parting of their ways, but it cost Craig $38 million to get Bill off his back. Cheap at the price, was the general consensus. TIGR pioneered the sequencing of bacterial genomes by shotgun. TIGR is now headed by Claire Fraser, previously Mrs. Craig Venter.

Gerry Rubin

I don't really care about human—mouse would be much more interesting (you can do experiments with mice). Human males die of prostate cancer and Congress is full of old white Anglo-Saxon men. Loads of money would go to the National Institutes of Health to sequence the genome of man. It would be a nice collaboration with the Wellcome Trust,[16] honoring an agreement forged on a windy February day in Bermuda to make all sequences public as they come off the machines.[17] Several good careers would be built sequencing the human genome bit by bit, clone by clone. Now Craig, always the maverick, comes

---

[16] The Wellcome Trust (WT) is a very well-endowed biomedical charity in the U.K. Funds the Sanger Institute and much university research. Without the WT, most biologists in the U.K. would have emigrated years ago. After Craig's announcement, it said that it would double its funding of the Sanger Centre (although, to be fair, that decision had been made months before). Nevertheless, the race between the public domain and Celera was on. Soon Incyte joined the fray (Marshall E. 1998. A second private genome project. *Science* 281: 1121), but..., oh, just but.

[17] The famous Bermuda Agreement (Dixon D.D. 1996. NIH seeks rapid sequence release. *Nature* 380: 279; Smith D. and Carrano A. 1996. International large-scale sequencing meeting. *Human Genome News* 7 [no. 6] April–June at http://www.ornl.gov/sci/techresources/Human_Genome/publicat/hgn/v7n6/19inetern.shtml) was on sequence data release. My notes taken at the meeting show that when Michael Morgan called for a vote, Craig raised his hand in favor of it, although he now denies that. It was, I would agree, a reluctant act and we all knew that his heart was not in it. The meeting was in Bermuda in February, but not for the obvious reason (although it happened that all accommodations on the British Virgin Islands at that time were booked). Bermuda is British, near the U.S., and offers good flight connections. As simple as that. For the controversy, see Bentley D.R. 1996. Genomic sequence information should be released immediately and freely into the public domain. *Science* 274: 533–534; Adams M.D. and Venter J.C. 1996. Should non-peer-reviewed raw DNA sequence data release be forced on the scientific community? *Science* 274: 5354–5356. Mark Adams and Craig's answer to this rhetorical question (by now this should come as no surprise to you) was "no." Note how even the title of this paper makes it clear that this is really for our own good: The data are not to be "forced" on us! There are many examples that one can give to illustrate the benefit of immediate public release; the only counterargument is greed, be it financial or professional.

John Sulston

along and says that this will never get done by gentlemanly science: It needs new machines, the business approach, and he will do this and both save the taxpayer dollars and make himself rich. And the elder Hunkapiller[18] has these new machines—ABI3700s, capillaries that can sequence 500,000 base pairs per day—and PE is setting Craig up with a new company, one that so far dares not speak its name, with 300 of them. That is 150 million base pairs a day. Awesome.[19]

Pity those who came to the Cold Spring Harbor Laboratory (CSHL) 1998 sequencing meeting to talk science. No one was interested in science. This was far more interesting. One can always rely on the fact that Craig's timing is perfect: Announce the commercial sequencing of human a day before the meeting and you have the world stage. Grace Auditorium is almost empty, at least empty of anyone who likes to think of himself (and they are mostly "him"selves) as a player. Small meetings in Jim's office, twos and threes walking down Bungtown Road oblivious to the spring warblers. Huddles in the basement bar in Blackford. Even small dinners in Huntington[20]—but maybe

---

[18] Michael Hunkapiller, here called the "elder" to distinguish him from his kid brother, Tim. M.H was a big honcho at Perkin-Elmer, a scientific equipment company that made and sold ABI machines—the workhorses of big sequencing centers. Tim is a bioinformaticist in Seattle, Washington.

[19] Marshall E. and Pennisi E. 1998. Hubris and the human genome. *Science* 280: 994–995.

[20] Grace is the auditorium at Cold Spring Harbor Laboratory in which big meetings are held. Replaced the small Bush lecture room, which was not air-conditioned, in 1986 (see Watson E.L. 1991. *Houses for science: A pictorial history of Cold Spring Harbor Laboratory.* Cold Spring Harbor Laboratory Press, Cold Spring Harbor, New York). Bungtown Road is a public road running through the Cold Spring Harbor Laboratory grounds. Blackford Hall is the Lab's dining facility, with a few guest rooms above and a bar below. I suppose one must, reluctantly, accept that the facilities improved after their renovation, but the food, alas, had not, although during prestigious meetings, it is *now* much better. Huntington is Cold Spring Harbor's nearest metropolis. A few good restaurants (Jonathan's can be recommended). The best sushi place is hard to find, and I am afraid that I won't help you with that one.

Francis Collins

that was simply to get some decent food, the CSHL kitchen being, what shall I say, "undistinguished." I do the only sensible thing and seek out Rob and Dick for a drink in the bar.[21]

But *Drosophila?* Why should Craig worry about the genome of a small fly? "Oh, he says that it will provide proof of principle for assembling human." Shotgun sequencing 3 billion base pairs is easy—we all know that. The hard part is assembling this sequence into anything remotely resembling the human genome's sequence.[22] An analogy may help: Take down a very large polychromatic brick building (the Natural History Museum on Cromwell Road would be ideal) and, as you do so, number each brick and make nice orderly piles of bricks. Given time and patience, you could rebuild Waterfield's masterpiece with little technical difficulty (this is what scientists call a gedanken experiment, one carried out in thought only). Repeat the experiment, but this time simply place the bricks, unnumbered, into random piles. Better yet, take 12 Natural History Museums and demolish them all, mixing their bricks.[23] The demolition will be quick and dirty, but rebuilding a single museum will be much harder. This is Craig's approach. That of the traditionalists is the orderly approach: Take one museum apart brick by brick, keep detailed records of the origin of every brick, and then rebuild.

So, the question arises again: "Can it be done?" In truth,

---

[21] Rob Martienssen, a plant geneticist at Cold Spring Harbor Laboratory. Had been my undergraduate student in Cambridge. A good drinking companion (and scientist). Dick McCombie worked in Craig's group at the NIH and is a great source for stories about Craig, as well as for suggestions for restaurants in Manhattan. Heads the sequencing lab at Cold Spring Harbor.

[22] Venter J.C., M.D., Sutton G.G., Kerlavage A.R., Smith H.O., Hunkapiller M., et al. 1998. Shotgun sequencing the human genome. *Science* 280: 1540–1542.

[23] I say 12, because a 12-fold coverage of the *Drosophila* genome, in which each of the approximately 175 million bases would, on average, be sequenced 12 times, would provide the assembly programs with a respectable chance of producing a good finished sequence.

JAMES D WATSON.   26 JUNE   CSHL   LEWIS MILLER 07

Jim Watson

nobody knows. For the genomes of bacteria, it certainly can—TIGR, Craig's "not for profit" in Rockville, Maryland, has proven this for several genomes. Gene[24] published a paper in 1997 in *Genome Research*[25] saying that it *could* be done for a genome as large as that of human. But this was immediately followed by a paper by Phil, the Phil of PHRED and PHRAP, that said "no way"—or, to be faithful to my sources, "In summary, clone-by-clone sequencing works and is cost-effective, neither of which appears likely for the whole-genome shotgun method of sequencing. There is no reason to switch."[26] In this trade, that is a strong put-down—not, for us, the histrionics of the *New York Review of Books*, at least not in print

"*Drosophila* is not a model for human." That is my only intellectual contribution to the debate. I know something about *Drosophila*, you see. I know that the pattern of repeated sequences (identical bricks, if you like) in *Drosophila* is quite different from that in human. And anyone with a modicum of knowledge knows that it will be these repeated sequences—enigmatic to the biologist as they are—that will make the assembly difficult.

But I am running ahead of myself in the story. Craig's shocking announcement came at a private meeting of the "Big Labs" the day before the Cold Spring Harbor meeting; I was not there of course. All I know is hearsay. Perkin-Elmer had designed a new sequencing machine—bigger, faster, better than anything we had ever seen. They committed $200 million to a new company to be headed by Craig to sequence and *sell* the human genome. There is that no-no word "sell": You can only sell

[24] Gene Myers, a software wizard and the real hero behind the Celera *Drosophila* sequence. Was then in the department of computer science at the University of Arizona, Tucson. Drives a Harley Davidson.

[25] Weber J.L. and Myers E.W. 1997. Human whole-genome shotgun sequencing. *Genome Res.* 7: 401–409.

[26] Green P. 1997. Against a whole-genome shotgun. *Genome Res.* 7: 410–417.

(legally) what you own. And how can anyone claim to own the human genome sequence? Does that not mean patent it?[27] Can we allow one company, one man—even a Vietnam vet—to own *our* genome? Surely that belongs to all human kind. "And as proof of principle, I will shotgun sequence the genome of a model organism in 2 weeks," announces Craig. Well, maybe I exaggerate the 2 weeks bit. Even so, whether it was 2 weeks or 3 months, it was plainly absurd. But what model genome? The worm was done (well, almost). *Arabidopsis*—a small weedy plant? What model? Gerry, of course, was there. He plays in the "Big League." After his talk about the mission of Celera, Craig introduces the elder Hunkapiller to describe the new ABI machines and walks out of the Plimpton conference room, passing Gerry on the way. "Can I have a word with you outside?" The penny drops. Of course, he means *Drosophila*, that little fly so loved by geneticists. Gerry is already sequencing *Drosophila* and Gerry, like me, has spent most of his professional life with this fly.

I still do not know why Craig wanted to sequence *Drosophila*—even now, when it is finished. Did he expect to make money from it? Surely not.[28] Nobody gets rich studying *Drosophila* (well, that may not be entirely true, but let's keep it as a first approximation). Or was the reason more mundane,

---

[27] The question of patenting human genes has been immensely controversial, both in the field and among the general public. It's too hard a problem to summarize here, but, broadly speaking, those of us who work in the public domain are against it. (See Poste G. 1995. The case for genomic patenting. *Nature* **376:** 523–526; Doll J.J. 1998. The patenting of DNA. *Science* **280:** 689–690; Heller M.A. and Eisenberg R.S. 1998. Can patents deter innovation? The anticommons in biomedical research. *Science* **280:** 698–701.)

[28] Of course, the fact that I do not understand the Celera business model speaks for nothing. I cannot think of a business model that I have ever really understood, and that has not stopped many of my contemporaries from becoming very rich. See Marshall E. 1999. A high-stakes gamble on genome sequencing. *Science* **284:** 1906–1909. This is worth a look for the photo of Craig in what appears to be a metal womb.

more pragmatic? By starting with an organism with a small genome, perhaps he could learn to do large-scale shotgun sequencing and assembly. Rather like running a 1000-meter race to see whether you can run a marathon.

Gerry is a master. How he keeps his head, I do not know. Against this man I will never play poker. There are mutterings that Gerry is collaborating with the Devil. "Remember Poland," says Jim, who recalls Word War II. Jim is strong on historical analogies this week: "Will you be Chamberlin or Churchill?" he asks Francis. Jim is an Anglophile.[29]

The world is agog. Of course, it is a pretty small world, but it is My World, the Golden Run from Cambridge to the east coast and on to Berkeley and Stanford. The e-mails begin to fly. Will Craig patent *Drosophila*? What are these new PE sequencing machines? Do they work? Why is Gerry collaborating with him? Will the sequence be public? When will it be done? Will I be scooped? What will this mean to the European *Drosophila* sequencing project? Paranoia. Pure paranoia fueled, as it is so often, by misinformation and lack of information. Gerry had his work cut out to convince the community.

The next stop is Crete, the Orthodox Academy at Kolymbari, just below the long finger of Diktynna pointing northward, west of Chania. For reasons well known to many—but hotly disputed by a few (well, one)—about 100 *Drosophila* biologists enjoy the hospitality of Alexandros Papaderos[30] and his monks[31] every 2 years to discuss who had done what to whom. But in this case, the whom is a fly. What surprises will be in store this year?

---

[29] I hope that the allusions in this paragraph do not need explanation.

[30] The director of the Orthodox Academy of Crete. Alexandros and his staff have welcomed drosophilists to Kolymbari every 2 years since 1978. A tireless worker for Cretan-German reconciliation, which is quite something when you consider what the Germans did to the Cretans, twice: first as Nazi invaders on May 20, 1941 and second as tourists.

[31] This is poetic license. Alexandros is not a monk. The monks are 100 meters up the track at the Ghonia Monastery.

What stories will be swapped swimming in the warm sea, or over wine and raki at Lefka Ori?[32] But it all comes down to the same thing: who is up and who is down. The European sequencers are not happy bunnies—minnows to the American sharks. But we put a good face on it. Yes we will collaborate, yes we will change our strategy, yes we will sequence the ends of our clones, big clones these, known as BACs (bacterial artificial chromosomes), made for us by our French collaborators. Formal agreements with Gerry, smiles all around, but tense, and Crete should not be tense. Crete is where drosophilists chill out, eat lamb chops with their fingers, go to mountain villages to eat mulberries, and drink ice-cold water off the White Mountains.[33] Crete is where drosophilists think and talk science, dance with chairs in their teeth, and break plates.[34] Not cut deals. But there you are: Extraordinary times demand sacrifice, even from drosophilists.

During the summer, a desultory "negotiation" with Celera takes place, principally about a FlyBase license. Celera wants

---

[32] Lefka Ori is a typical Greek taverna at the far end of the village, near the main road to Chania (on the right, as you turn off the main road). It is traditional for many attendees of the Crete meeting to have an evening there. Nikos Vrouvakis is famous for his hospitality. Say that you are a friend of Fotis or Spyros.

[33] I am afraid that the name of this village, to which a group of us, including Peter Lawrence, Gines Morata, and Pat Simpson, walk every time we are in Kolymbari, must remain a secret.

[34] I must admit to having only seen dancing with a chair between the teeth once (it was Kitsos Louis, from the Institute of Molecular Biology and Biotechnology in Heraklion). But that is the evening that every plate in the taverna was broken and two very distinguished Greek biologists of the diaspora drove me back to the Academy (all of one mile) as drunk as coots (*Fulica atra*; the North American species is *Fulica americana*, but they look and behave much the same). We then all went for a swim at 3 a.m. to sober up. It did not work. The extraordinary sequel was that Spyros Artavanis-Tsakonas managed to find money in the meeting budget to pay Nikos the cost of the damage.

FlyBase.[35] The problem is that we are dealing with the suits—the "Director of Business Development," no less. At one point (later, in September), MarkA[36] says to me, "I will do everything possible to make life difficult for FlyBase." Cheery. Bill threatens to go into "omerta" mode (and him, with not a drop of Chianti in his veins).[37] We are out of our depth, I feel. Other sharks are in the water, too. Mark Patterson[38] went to see Craig on a grand tour in July to sweet-talk him into publishing the *Drosophila* papers in *Nature*. He even offered to organize a "press conference" to mark them. As I said, we are out of our depth.

By September, it is Miami Beach, with the extraordinary chrome and stained glass of the Fontainebleau Hilton and the ghost of Sinatra haunting its ballroom (or was that my imagination?). This is Craig's meeting—GSAC9. Even Clinton is rumored to be making an appearance. The logo of Craig's new company, Celera Genomics, faces you at every turn. This is the Big Time. PE rolled out the ABI3700 and promoted as if it were this year's new Mercedes complete with dolly birds. And this is science? Of course, nobody knows whether it will work or can answer the critical question of the read length.[39] Will it be 400

---

[35] FlyBase (http://flybase.bio.indiana.edu/) is the database for drosophilists. Started in 1992, it is funded by the NIH and the Medical Research Council in London. Bill Gelbart (see footnote 37) is the principle investigator (PI) for FlyBase. Gerry and I, with Thom Kaufman and Kathy Matthews in Bloomington, Indiana, are co-PIs.

[36] Mark Adams, a long-time colleague of Craig's, and then a vice president at Celera.

[37] William Gelbart, known as Bill except by his family, who call him Billy. Has a good intellectual pedigree: Ph.D. with Art Chovnick and postdoctoral fellow with Ed Lewis (our hero) at Caltech. Discovered *decapentaplegic*, an unbelievably ugly mutation in *Drosophila* (see the cover of *Cell*, March 1982). That turned out to be important. Now a professor at Harvard, Bill is from the Bronx. Or is it Queens? All I know about Queens is that you keep your head down when driving through it from Manhattan to J.F.K. Airport.

[38] A *Nature* editor who had been an undergraduate student of mine and still lives in Cambridge.

[39] The read length is the number of base pairs that can be distinguished in a single sequencing run.

base pairs per lane? If so, Craig is in deep shit; 500?—dodgy; 600 or more?—laughing all the way to the bank.[40] Craig has made a smart move. He has hired the euphoniously named Gene Myers to write the assembler, the computer code that will build the new museum from the tons of unordered bricks. But then, nobody ever said that Craig was stupid.

I hate these hotels, loathe them, abhor them. The windows never open and the coffee is undrinkable. And it is raining, with Hurricane Frances beginning to blow in. But the worst part is meetings in windowless rooms around round tables, tables full of jugs of stale water. We meet with Craig. Gerry and I discuss various scenarios beforehand. Gerry thinks Craig will deliver: He wants to be able to say "I sequenced *Drosophila*." But he has already tied up two pharmas and they will want privileged access.[41] Craig tells Gerry, "Do not do anything at this stage which would be incompatible with finishing the fly genome yourself." Adam,[42] Gerry's NIH grant manager, nods in agreement. Will Craig *really* make the data freely available to the community, or will he try to license it, so that the big bad boys in Incyte cannot suck it up and sell it on? All Craig says is that "We are still trying to work out how to do it." Not very satisfying. Richard[43] is not happy, and an unhappy Aussie is not a pretty sight. I get paranoid when I later learn that Gerry and Craig had a meeting before Miami. Does Gerry have a private agenda?

In October, we take advantage of Gerry's seminar at the European Molecular Biology Laboratory (EMBL) in Heidelberg

---

[40] The longer the read lengths, the easier for the assembly programs to build a finished sequence. Long read lengths are also important because they reduce the number of sequence runs, and hence the time, needed to sequence the genome.

[41] Indeed, they get it.

[42] Adam Felsenfeld, NIH grant administrator.

[43] Richard Gibbs, head of the sequencing center (one of the "Big Five") at Baylor College of Medicine. Collaborated with Gerry on sequencing *Drosophila*.

to have another meeting. Gerry, Fotis, David,[44] and I sit in Fotis' newly renovated office. (What did that cost the EMBL? I thought that we were broke.) We mostly discuss BAC strategies. The problem for us in Europe is simple: no money. All very frustrating. Gerry discusses the pressures on Craig from his investors, from his customers, from his Scientific Advisory Board, from the community. Can we use these? We go over the same ground later that month when the European *Drosophila* Genome Project (EDGP) meets in Lisbon. Gerry absolutely refuses to join us in Lisbon. (Not that I blame him for that. The Delta flight from Lisbon to J.F.K. is the pits.) We want to be involved, but how? David and I come up with some money to pay Jean[45] to sequence our BAC clone ends at Genoscope.[46]

By October, the *scélérates*[47] have a home in Gaithersburg, Maryland. At least it's only 40 minutes from Dulles Airport. This time, Gerry cannot join Bill, Suzi,[48] and I for what, ludicrously, my premeeting note said is "not a negotiating meeting." We are given

---

[44] Fotis Kafatos was one of my bosses. Is Cretan and a member of the Orthodox Academy of Crete and many other academies. Works on the malaria mosquito these days. Was then Director-General of the EMBL, headquartered in Heidelberg. Hence, my and Graham Cameron's hatred of Frankfurt airport (see below). David Glover, now professor of genetics in Cambridge but then at the University of Dundee. Coordinator of the European *Drosophila* Genome Project.

[45] Jean Weissenbach, head of the major French DNA sequencing lab, Genoscope, at Evry. You pass Evry driving on the A6 to Nice, but if you have any sense, you keep your eye on the road and try to make Le Marche in Beaune in time for dinner (but book first).

[46] BACs, bacterial artificial chromosomes, are long DNA clones. When the Berkeley and European sequencing projects started, they used relatively small cloned pieces of DNA to sequence: either P1 clones or cosmids. BAC clones, being much longer, were a great improvement. By sequencing the ends of BACs, we obtain paired sequences of great utility for genome assembly.

[47] "An atrociously wicked person, a villain, wretch." *Oxford English Dictionary*, 1933.

[48] Suzanna Lewis was head of Gerry's informatics group at Berkeley. Trained as a biologist, but circumstances lead her to computing. A heroine of this story.

a tour (this is where the sequencing machines will be) of a huge empty shell of a room (this is where we will have the biggest and best and fastest computers outside the National Security Agency; this is where the power plant will be). Bare pipes and concrete everywhere. The only habitable rooms are Craig's office (this guy has a positive self-image; his smiling face can be seen on framed covers of *Time*, *Newsweek*, and *The Gaithersburg Weekly Herald*) and a conference room. And he says that he will sequence 2 billion base pairs of *Drosophila* by April? Not a sequencing machine in sight. Crazy. "Craig, we can only collaborate with you if the data goes into the public domain." "No problem [the suit across the table nods in agreement]; I will deposit it in GenBank,[49] but with a notice saying that it cannot be redistributed or sold." "No, that is not possible. Data in GenBank cannot be restricted." The arguments get tedious. I repeatedly walk out for a smoke. "I will license the data to FlyBase for distribution if FlyBase licenses the *Drosophila* database to Celera." Why does Craig want our database? If we give it to him, will he sell it to his customers? Will he put us out of business? But FlyBase is also publicly funded. Yes, U.S. taxpayers pay for a database on *Drosophila*. In our eagerness to see the data public, do Suzi, Bill, and I go too far? Dammit, I was trained to push flies, not negotiate complex agreements. About the only item on which we agree is that Celera will not poach FlyBase staff, and vice versa, but even Bill said that with tongue-in-cheek.

Back to Celera in December. Gerry and I had been at a Banbury[50] meeting but did not travel together. We never do. Craig reviews the timetable for sequencing the fly. Finish by June 1, 1999. If Gerry collaborates on gap closure in a "balls-to-the-wall" effort, then Celera will deposit the sequences in GenBank by July 1. If not, then on publication. More tedious discussion

---

[49] GenBank is one of the three databases that make up the international nucleotide sequence data bank; the other two are the EMBL Nucleotide Sequence Data Library, produced at the European Bioinformatics Institute at Hinxton, and the DNA Data Bank of Japan, produced in Mishima.

[50] The Banbury Conference Center of Cold Spring Harbor Laboratory is one of the very nicest places that I know for a small meeting.

about what GenBank is and what it is not. Surely Craig already knows this. Further discussion of a possible deal of a FlyBase license for Celera in return for the data. The drive that Sunday evening (Gerry, Bill, and I) to some godforsaken golf resort in Virginia[51] (why does the NIH do this to me? Yes, I take it very personally) is tense. I am relieved only by the sight of a flock of starlings[52] that must have included in excess of a million birds. I don't think Bill or Gerry notice. It is no surprise that Bill and I head straight for the bar. Bill is addicted to margaritas (with ice and salt), and I am desperate for a scotch—but Gerry? No, I should not have been surprised, for it had been a tense day, that Gerry joins me in a scotch (or two)[53] The dilemma is obvious. If we can bring this off, we will all benefit, and by "we" I mean not we three but the scientific community in general. Does that sound pompous? It may, but we meant it. If it is screwed up then we, and here I do mean "we three," will be blamed. Dammit, even the sequence of that little worm is now finished—it is to be published in a couple of weeks. There is really only one question: Can we trust Craig? If so, can we forge a deal that will satisfy the community's need for free access to the data?

Gerry should have been at the NIH workshop on Monday but is not. He returns, smiling, at what I am trying to convince myself is tea time. "It's OK. I had lunch with Harold[54] and Francis" (big honchos at the NIH, they pay the bills in Gerry's sequencing lab). "If Craig agrees to the data going into GenBank and as long as any data I contribute go into GenBank immediately then we can collaborate." This *is* news—the mutual animosity between the NIH and Craig is well known and goes back years. But Gerry, Gerry

---

[51] The Lansdowne Resort. We were on our way to a NIH meeting, whose purpose was never very clear to me. It seemed as if it was to tell the rat community how to write a grant proposal for a rat genetic database.

[52] *Sturnus vulgaris*, yet another European contribution to the U.S.A.

[53] I do not wish to imply that Gerry is a teetotaller, but that he drinks with due moderation.

[54] Harold Varmus, director of the National Institutes of Health (until December 1999). Widely considered to have been an outstanding success.

from Berkeley by way of MIT, Cambridge, Stanford, and Harvard Medical School,[55] has pulled it off. I write a note to myself: "I predict that there is a 60% chance that Craig will abort." But I was wrong. I had underestimated the extent to which Celera really needed the Berkeley data to do a decent job. Within weeks, there are smiles all around as both Francis and Craig announce a joint "Memorandum of Understanding" (MOU)—good press for both.[56] "The early availability of the *Drosophila* genome is a meritorious goal." And yes, there is the commitment: "Celera Genomics will begin depositing contigs, as unfinished sequence, in GenBank upon the completion of the random sequencing phase." Signed and sealed. Not legally binding, of course; either side can withdraw from the agreement with 30 days notice. But pretty good, although Celera cannot resist twisting the knife: "Celera Genomics, at no cost to the federal government, will independently accomplish the following tasks."

Are people happy? Of course not. I could hear them already: Gerry has links with a company, he will let them see the data first, make loads of money, and beat me to the discovery of the secret of life. Craig will welsh on the deal, he will patent everything, and we won't be able to do any experiments without paying him money. Novartis[57] has made a deal with Celera and Bill will consult with Novartis; he will steal a march on understanding the *decapentaplegic* pathway ("what?"—don't worry about it). Pharmacia & Upjohn[58]

---

[55] Gerry was an undergraduate at MIT, did his Ph.D. in the Medical Research Council Laboratory for Molecular Biology in Cambridge (in the good old days), was a postdoctoral fellow in Stanford with David Hogness (the guy who really founded the science of genomics), and was a junior faculty member of Harvard Medical School. A good flavor of Gerry's negotiating skills can be had from an account of his recruitment by Berkeley. This was told to L.M. Rubin and privately printed in a special "issue" of *Science* dedicated to Dan Koshland, then editor of that journal, on the occasion of his 70th birthday. A copy can be found in the CSHL Archives.

[56] See News in Brief 1999. *Nature* **397**: 380–381; Pennisi E. 1999. Fruitfly researchers sign pact with Celera. *Science* **283**: 767.

[57] A big pharma; Swiss.

[58] Another big pharma. Pharmacia was Swedish and Upjohn American, so I guess that P&U was Swedo-Amerish.

has made a deal with Celera; I will consult with them and steal all of the genes controlling sex and death. Pharmacia & Upjohn has made a deal with Exelixis (Gerry's company[59]). The circle is closed. We are just in it for the bucks. Total paranoia. Thank God that I will miss the annual meeting of drosophilists in Seattle this year.[60] FlyBase takes me there beforehand and it turns out to be another godforsaken hotel in the wasteland of the most inappropriately named Belleview. I was getting tired of shouting at Bill and he, I am very sure, was getting tired of being shouted at by me. Sorry Billy. I escape back to Europe (upgraded this time; I deserve it) just as everyone else is checking in for the meeting. Bill and Gerry can take the heat. Luckily, it is raining.

It is now March (1999, if you are lost). Didn't Craig say that he would finish the fly sequence by April? He has not even started yet. I told you that he was full of bullshit. Ha! ha! He cannot do it. Relax; he will—give him time. Who are these people one has as colleagues? Hobby sequencers. When is the sequence going to be finished? When can I have it? Why is it not released yet? How many genes are there? Where is my gene? Paranoia rules all summer. I am only relieved by a trip to Iceland[61] and a visit with that other maverick of genetics, Kari of the Saga and deCODE,[62] a start-up company that employs 0.1% of the pop-

---

[59] Exelixis (Greek for "evolution") was the brainchild of Spyros Artavanis Tsakonas. Spyros is now at Harvard Medical School, perhaps en route to Paris, having escaped a long sentence at Yale. He, Gerry, and Cory Goodman (a colleague of Gerry's at Berkeley) founded this company to make money. Gerry was, at the time, only a member of the company's Scientific Advisory Board. He has now resigned, for reasons that may become apparent (if they do not, the reasons were wholly honorable).

[60] The world's drosophilists, at least 1500 of them, meet in confero every spring. In the good old days, we met on a university campus or, better, at the California State Park's conference center at Asilomar. Now there are too many of us and we have to go to grotty hotels.

[61] I was on vacation, birdwatching and not thinking about flies.

[62] Kari Steffanson was a professor at Harvard Medical School. He had the bright idea of using the genealogical and health records of Iceland to map genes responsible for human disease or susceptibility to disease. This has been very controversial, but despite what some people write (e.g., see the letters to the edi-

ulation of Iceland. (It is not only Craig who thinks big.)

It is now getting serious. Rumors fly. Craig has no intention of releasing the data to GenBank. He will make a deal with the National Center for Biotechnology Information (NCBI) and make it available only under all sorts of restriction. "Francis, I must talk to you." But the BBC Money Programme[63] also wants to talk to Francis, Jim, Michael Morgan, and all of the other Big Shots on the lawn at Airsley.[64] Yes, 12 months later we are back at the CSHL sequencing meeting.[65] "Francis, I think this is important." That is the problem with being short: Getting the attention of a tall man means that one has to resort to histrionics. "Francis, I am hearing all sorts of rumors, from usually reliable sources, that Craig is going to break the MOU with Gerry and licence the data." Well, you do not get to be head of the National Institute for Human Genome Research without the ability to calm excitable English gentlemen. But I am not convinced. A long e-mail to Francis on my return to Cambridge means, at least, that I can say "I told you so" if the shit really hits the fan.

---

tor in *Nature*, 19 August 1999) it has the general support of the Icelandic population (see Enserink M. 1998. Opponents criticize Iceland's database. *Science* 282: 859; Abbott A. 2000. Iceland's doctors rebuffed in health data row. *Nature* 406: 819). Kari phoned me very late one night (I was asleep) at Mývaten in northern Iceland. How he knew that I was there I never discovered (although his father was a member of Parliament for the region—I hope that he liked sheep). Anyway, Kari demanded my presence in Reykjavik. Of course I went, although I never had the nerve to send him a bill for consultancy. Kari's brother Haldor works for the EMBL in Heidelberg and Haldor's wife, Anne Ephrussi, is a well-known drosophilist who had worked in Harvard BioLabs as an undergraduate. Her father Boris was a member of the Caltech *Drosophila* mafia, but was unfaithful on his return to France and worked on yeast. Small world.

[63] Shown on BBC2 on June 13, 1999.

[64] The president's house at Cold Spring Harbor Laboratory. Jim used to live there when he was president, but as chancellor now has a large Palladian mansion (see footnote 20). The president (now Bruce Stillman) always throws a garden party during meetings at the Lab.

[65] At which the Big Three public labs all put up speakers to trash the ABI3700; see also Mullikin J.C. and McMurray A.A. 1999. Sequencing the genome, fast. *Science* 283: 1867–1868.

2

The Middle Game

Gerry again. Two rules: (1) Never do anything without a back-up plan. (2) Keep smiling in public. Not explicitly stated, but I can still learn, even at my age. OK, let's begin to plan. We need to plan what we will do with the sequence once it arrives in our computers. What does a DNA sequence look like? At best, like this: ATGGTATTAAGTCCA... . I can read that: methionine, valine, leucine, serine, and proline. The first amino acids of the globin alpha chain, the protein that carries oxygen in our blood. At worst, like this: AAAAAAAATTAATAAATATAAAATTTTTT, which does not seem to code for anything. That was a simple one. But if there are 175 million of these As, Ts, Cs, and Gs, we have a problem. Some stretches will code for proteins, the stuff of living cells; other stretches will have the information for controlling the readout of the genetic code; other stretches will be junk, or worse, garbage. (Junk is what you keep to reuse on a rainy day. Garbage is what you throw away. This definition is Uncle Syd's.[66] It is becoming fashionable for the scientific equivalent of the chattering classes to quote Uncle Syd. For example, those of us who believe in the power of clever algorithms and clean computer programs to make sense of the As, Ts, Gs, and Cs he calls biofanaticists. Syd, that is not up to your usual stan-

---

[66] Sydney Brenner. What can I say about Sydney? That he refused to take me as a graduate student? Or that he is one of the most brilliant molecular biologists of his generation? Read his book, *Loose Ends* (Current Biology Ltd., London, 1997). The person referred to on page 61 is, by the way, Ben Lewin, founder and long-time owner/editor of *Cell*.

dard of puns.[67] Living in the Shattuck Hotel in Berkeley must be getting you down [now, *that* is a hotel I can recommend: large airy rooms with 1920s furniture. But please do not try and reserve the corner suite on the third floor; that's mine. And it is only 45 minutes from Point Reyes[68]]).

Where was I? Planning. How can we analyze the sequence of *Drosophila*? Someone, either Gerry or Suzi, had a great idea. Let's get a bunch of people together for 2 weeks to sit in front of a group of computers and just go for it. Write some cool Perl, get Neomorphic Software, Inc. to write some tools,[69] and we could be in the best of both worlds—have fun and do some neat science at the same time. There must be a catch. Well, it will have to be at Celera's factory and that means 2 weeks in Gaithersburg, Maryland. Will Celera buy in? Will they? Of course they will. They see a good deal when it is offered on a plate. For the price of a contract with Neo to develop the tools and for putting up 40 or so people for a couple of weeks, they get a consultancy that money could not otherwise buy. The "jamboree" is born. Suzi comes from a scouting family and is

---

[67] I quite like his statement that "Ontology recapitulates philology" (Brenner S. 2002. Life sentences: Ontology recapitulates philology. *Genome Biol.* **3**: Comment 1006. 1–2), although I think that he missed the point and he upset some of my Gene Ontology colleagues. Much later, at the 2005 CSHL Symposium, Sydney referred to the Human Genome Project (HGP) as being "Stalinist," to which Eric Lander (after, I should add, Sydney had left the room) replied "Stalinist? But we had weekly conference calls." That is one small illustration of the fact that this story is, by comparison with the HGP, a nursery tale. It is also a lesson in the subjectivity of history.

[68] A great place north of San Francisco. Take the Golden Gate Bridge out of the city and drive up Highway 1. Good for sea birds and whales. Stop for oysters at Inverness (the place on the left, driving south) on the way back.

[69] How grand that company sounds—founded by a bunch of bright kids from Gerry's lab, who wrote nice tools. Open kimono time: I had an option in their stock.

PORTRAIT OF SUZI LEWIS   IS OUT ISHA (DEVENPORT)   LEWIS MILLER 05

Suzi Lewis

obviously nostalgic for the long summers on the UP[70]—hence the term "jamboree." I don't object, even though my father forbade me to join the scouts on the grounds that they were a paramilitary organization. He was right, but it took me three weeks as a scout to believe him. It was bad enough being the only Jew in the school without also being the only non-scout. But I digress to the point of self-indulgence.

If it is August (yes, still 1999), it must be the Stadthalle in Heidelberg, site of ISMB '99 (Intelligent Systems in Molecular Biology) (I am beginning to think that Uncle Syd may have a point, but at least the Santa Fairies[71] stay away). We were there because we were holding a bake-off. It isn't enough to sequence a genome, it has to be analyzed—the technical word is annotate—to figure out what's there. As a practice run for the whole genome (Craig isn't the only one who believes in rehearsals), we had completed the sequencing and analysis of three megabases of *Drosophila*. This particular region was chosen because it surrounds a gene near and dear to my heart, the *Adh* gene, which prevents flies from getting drunk.[72] We had used state-of-the-art software for the analysis, but then went the extra mile and followed these suggestions with wet-bench work to be certain about what was there. Martin,[73] one of Neo's founders and Gerry's graduate student, had the bright idea of asking as many ambitious young souls as possible to take a crack at analyzing

---

[70] The Upper Peninsula, Michigan. The nearest I have been is Ann Arbor (pretty grim, and you must use Wayne State Airport in Detroit, which holds my prize for being the very worst airport in the U.S.). Sorry, I have no recommendations for the UP. Ask Suzi.

[71] A bad pun. Think of Stuart Kauffman.

[72] The story once went the rounds that, for this reason, my work was funded by Greene King, the great East Anglian brewers. Far from it; we contributed generously to their profits.

[73] Martin Reese, onetime German national student table tennis champion, was a founder of Neomorphic. Martin has a very bouncy and optimistic personality, along the lines of Tigger.

PORTRAIT OF MARTIN REESE   4 NOV CSHL (DAVENPORT)   LEWIS MILLER 05

Martin Reese

the same sequence to see how they compared (much in the Tom Sawyer mode). We were presenting the results of these comparisons in Heidelberg, a first GASP (genome annotation assessment project), but this was just the public story.[74]

In Heidelberg from Celera are Peter, Tony, Gene, and MarkY ("MarkY" is used to distinguish him from MarkA, MarkY's boss at Celera).[75] From the public side are Suzi, Martin, and me. Luckily, it's August, which in Heidelberg means the season of pfifferling.[76] The business, of course, is over food. The sequencing is going well, we hear. They will be finished with the sequence in September. Does the assembler work? A smile from Gene, whose team of software engineers has been busy. (What we were not told then is that when they first ran the code, they got lots of nonsense—it took them 4 days to find the one line of buggy code. I hope Gene sweats sweet.) Gene presents his simulation experiments at ISMB. It should work, but you cannot convince a fly pusher with a computer simulation; however, the computer jocks are impressed. But Celera still wants a FlyBase license. Negotiations with Bill (because Bill is the PI on FlyBase, he is the boss) have not been going well (I think that may be my first deliberate understatement). Bill is pissed off because Craig will not return his calls. Craig gets his suits to call, and Bill hates lawyers. Worse, Bill has told his FlyBase peons that they must not

---

[74] For some time, I managed to raise a weak laugh during seminars by saying that this was also the last GASP. But I can do so no longer, because a replay was done for the human genome annotation at Hinxton in the spring of 2005. See Abbott A. 2005. Competition boosts bid to find human genes. *Nature* 435: 134; Guigó R. and Reese M.G. 2005. EGASP: Collaboration through competition to find human genes. *Nat. Methods* 2. 575–577.

[75] Peter Li was with the Human Genome Database at Johns Hopkins, but when that went belly up, he joined Celera to run the chromosome team. An unusual man; a medic who writes Perl and shell scripts and has a quite insatiable passion for sushi (see below). Tony Kerlavage was vice president of Celera and head of the gene discovery group. Mark Yandell was a member of the Celera chromosome team and primarily responsible for the annotation pipeline.

[76] *Cantharellus cibarius*, otherwise the chanterelle.

talk to Celera, that only he can talk to them. But although he can, he won't. Stalemate. It has become automatic for me to look at my watch and calculate the time in the other Cambridge and in Berkeley. A glass of Weissbier and a mobile phone. "Bill, it's August, it's Heidelberg, Tony is here, and I have to talk to him. Can I deal?" Bill is not happy, but he knows that it would be silly for me to disappear behind a pillar every time I see Tony. Anyway, Tony has a Celera Black American Express card[77] and I know some very good restaurants in town; I can take him to have pfifferling at Der Plock[78] or even Simplicissimus. Tony is reasonable (of course he is reasonable). Yes, Tony will have a face to face with Bill if necessary. Yes, he will go to Harvard. Yes, of course we can come to some agreement. Yes, let's keep the lawyers out of it. Yes, Yes, Yes. "Bill, it's fixed. Tony will call you." Well, it wasn't. The details are boring. I got tense. Bill got tense. Craig, I am reliably informed, called us all sons of bitches. Even Gerry got tense. But let us not go over it all again. Recriminations are never pleasant and rarely productive: In the end, it *was* settled. Celera got a free license to FlyBase (did I hear you say that they are taxpayers, too? I guess their stockholders live in hope).

Meanwhile, Suzi and I had real work to do. Just how were we to organize the annotation of the genome? We had to be prepared. We had agreed on a jamboree, but not much else. Had Celera considered the logistics? We would need T-shirts,[79] an expresso machine (no way was I going to spend 2 weeks in Gaithersburg, Maryland otherwise), a crèche in case kids came

---

[77] I am making this up, but if he doesn't, he should.

[78] Now, sadly, closed. Was the favorite restaurant of Fotis Kafatos. Apart from its food, this restaurant was distinguished by being run by a lady whom I can only describe as a dragon. It had a special vomit cup in the male lavatory. All I can deduce is that the aim of the average Heidelberg undergraduate must have been much better than that of his contemporaries in Cambridge.

[79] For the jamboree, Suzi bought a job lot of gothic T-shirts from Rasputin's on Telegraph Avenue in Berkeley and gave these out as prizes—for the best of whatever tickled her fancy that day, perhaps the best discovery—during the evening bull sessions (see below).

along, computers, software, telephones, fax machines; we would need hotel rooms, limos, and take-away sushi. Our demands were (almost) endless. Peter diligently wrote notes over lunch. The Celera staff would have to be trained. Only two people in the world had annotated large amounts of *Drosophila* sequence—Sima, working on the *Adh* region, and Takis,[80] who had annotated the sequence from the European project—and we thought that Sima's newborn would keep her in Berkeley. That left Takis. Martin wanted to run Genie—the best gene prediction computer program in the world (these are Martin's words, not mine). Would Celera send the sequence to Berkeley so that they could do this there on their computers? We needed to know how to move data around. Data are not something you can carry in your briefcase, they must be packed inside a special computer code; Suzi was pushing for XML, in particular for something she was inventing called GAME-XML.[81] Never ask Suzi what GAME stands for; she has forgotten. It is something along the lines of "gene annotation markup environment." We must have the DTD for GAME finished, Suzi insisted to MarkY. "Yep" (MarkY is from East Texas). (It is a measure of my tolerance of human nature that I can actually have dinner with people who speak in acronyms; this is, I think, the most corrupting influence on my character that I can place at the door of the

---

[80] Sima Misra worked with Gerry in Berkeley and was primarily responsible for the annotation of 3 million base pairs of *Drosophila* sequence (see Ashburner M., Misra S., Roote J., Lewis S.E., Blazej R., Davis T., et al. 1999. An exploration of the sequence of a 2.9-mb region of the genome of *Drosophila melanogaster:* The *Adh* region. *Genetics* 153: 179–219). Sima had her second child, Aden Misra Siebel, in February 1999 and it was touch and go regarding which event would happen first—finishing the annotation or the birth of the baby. Aden waited, bless him. Takis Benos or, as he likes to style himself, the Reverend P. Benos (an odd thing to do, especially for a Greek), was my postdoctoral fellow at the European Bioinformatics Institute, where he annotated 2.5 mb of X chromosome sequence from the European *Drosophila* Genome Project. Smokes a pipe.

[81] Much, I am sure, to the displeasure of her family, Suzi had done this during her vacation at Salmon Lake, where she camps with her family every year.

European Commission.) We left Heidelberg quite happy. I even managed to get the early Luftwaffe to Stansted[82] and be back in my office for the eclipse.[83] Suzi was happy because MarkY liked GAME. Martin was happy, because he was in his native Germany (and not California) and because the GASP workshop had gone well. I was happy because I thought that I had brokered a deal between FlyBase and Celera and that the jamboree might really happen. I hoped the Celera-ites were happy and that the suits did not hassle them over their expenses.

When I was young, I sneered at those who every June went to Cowes to yacht, to Henley to row, to Ascot to race, and every August to the Highlands of Scotland to slaughter grouse. What a boring life. And where was I this September 1999? Yes, the Fontainebleau Hilton on Miami Beach. At GSAC10, narrowly having escaped from Hurricane Floyd. MarkB[84] was there. How nice. He had been the graduate student who started DNA sequencing in my lab in 1979 and was now European president, no less (but no more), of Incyte.[85] Despite having moved to Cambridge 4 years ago and working just 100 meters

---

[82] This flight leaves very early. But at least it meant that you did not have to fly KLM UK (known locally as KLM-DELAY). It does leave from terminal 1 in Frankfurt, which is not an advertisement to German efficiency. Frankly, it is a mess. Even Graham (see footnote 92) got lost there once, and he used to go through it every week when he was setting up the EBI (that year, even Graham made Royal Sovereign on Air UK, as it was then called).

[83] There was a complete eclipse of the sun on August 11. Many from the ISMB went south to see it. Suzi gave up her ticket for the trip to a good viewing location and flew back to San Francisco. Those who did go enjoyed the sponsorship of Celera's European competitor, Lion Bioscience, but the clouds allowed only a brief glimpse of the eclipse.

[84] Mark Bodmer. I think that he left Incyte because they expected him to fly London–San Francisco every month, cattle class. M.B. had a British Airways Gold Executive card. His dad is a big cheese in genetics and was director of the Imperial Cancer Research Fund.

[85] A genomics company, great rival of Celera; their headquarters are in Palo Alto, California. If you kept your ears open at breakfast at the Sheraton in Palo Alto, you might have learned some commercial secrets. Bought out

from my house, I meet him for the first time in 3 years in a hotel lobby in Miami Beach.

Drosophila is done. Celera tells the world in a press release on September 9.[86] "Our team has done a great job," says Craig. So Sunday is a big day—the *Drosophila* day. Celera will talk about their fly sequence in public; Gerry will talk, and even I will talk about a mere 3 million base pairs that Gerry had sequenced and that we had jointly analyzed (that was what Sima had done and what we had presented at GASP). All we need is an audience. Despite the fact that 2000 people attend, most are enjoying the commercial freebies in the exhibition area— women with barely covered pudenda selling all sorts of gizmos and giving away balloons, T-shirts, mugs, and such a variety of useless paraphernalia that I cannot see my bed. Neo attempted to be the coolest kid on the block with fridge magnets showing their famous genome browser. Craig had a special *Drosophila* T-shirt made, with a nice picture of a sequencing reaction. "I see that you have made a few base pairs of your sequence public," I tell MarkA. "Not at all. That is Berkeley sequence." I am not joking; the suits are in control.

Now for the evidence: MarkA and Gene do a Box and Cox.[87] Gene gives the outline of the assembly program, MarkA of the sequencing itself, and Gene then provides the vital quality-control data. It's one thing to assemble the sequence, but how do you know that it's correct? After all, it has never been done

---

Hexagen, a company headed by Bodmer and founded by Peter Goodfellow and his colleagues from my Cambridge department. Having done that, Peter went off to work for SmithKline Beecham, a big pharma. At GSAC10, Incyte made a big splash—even the balls in the swimming pool and the door cards were decorated with their logo, and they had a TV program every morning. I never watched, but wish now that I had given them an interview about the young Bodmer.

[86] http://www.pecorporation.com/press/prccorp0909a.html

[87] The allusion is to an operetta by Arthur Sullivan and libretto by F.C. Burnand. Sorry if you did not know this. Check it out.

PORTRAIT OF GENE MYERS   CSRL 31 OLS (DAVENPORT)   LEWIS MILLER '05

Gene Myers

before. Well no, it hasn't, but Gerry is not just a pretty face. His lab at LBL[88] had sequenced 22 million base pairs already and Gene could compare this with his assembly. It was perfect; well, nearly perfect. One glitch: "We do not know whether our sequence is wrong or whether the Berkeley sequence is wrong." Gene says that with a smile. We all know what he means—it is not he who is wrong. Gerry hates that. He must have gotten either Sue[89] or Suzi out of their beds on Sunday morning in Berkeley to check this and check that. Within hours, Gerry can report that the one glitch is due to a Berkeley clone already labeled in the database as being a chimera—a wrong un. Had Gene checked the annotation, he would never have used it. "Jesus, the bastards have done it" (I overheard that in the bar. I think the speaker worked for a company in Palo Alto). They had indeed. The cynics were wrong. The sceptics should eat their hats in public (but they won't, and not just because of a fear of mercury poisoning[90]).

It was time to shit or get off the pot. Back to Celera: Would you believe that the UA[91] flight from Miami to Dulles left early? Life was really looking up. Serious discussions were underway with MarkY and Peter on just how we were going to analyze this beast. Suzi was worried that none of the preparations we need-ed seemed to have been done. Were they taking this seriously?

---

[88] Lawrence Berkeley National Laboratory. Funded by the U.S. Department of Energy, which older readers may know as the U.S. Atomic Energy Commission. These are the guys who blew up Bikini Atoll and worse. They are now trying to make amends by doing biology.

[89] Sue Celniker, head of the *Drosophila* sequencing group at Lawrence Berkeley Laboratory and the person who now has the job of "closing" the "complete" *Drosophila* sequence.

[90] Hence, Lewis Carroll's Mad Hatter. I strongly recommend the edition of *Alice in Wonderland* edited by Peter Heath, *The philosopher's Alice*. Academy Editions, London, 1974. Nobody in this business should fail to read this.

[91] United Airlines. I think that they now no longer have the nerve to tell us to "fly the friendly skies of United," but I may be wrong.

What precomputes were they going to do? Had the Celera annotators been trained? What software tools were going to be available? How were we going to orchestrate 80 people to work together on a common project? What kind of training with the software tools would be available? Worse, we did not even have a date fixed yet. One fact you must know about Peter and MarkY is that they work best over sushi. I had, once, embarrassed myself in Mishima in a sushi bar by being greedy; Graham knows the story.[92] That day with Peter and MarkY I learned that I was but an amateur, a beginner in the sushi-eating stakes. Even Suzi was impressed and she can be a hard gal to impress at times. (The lady who owns Sushi Ninja welcomed us as old friends 2 months later. I gather that that first session is now a legend among the sushi-bar owners of Rockville, Maryland.) Well, over the uni, ikura, and unagi—I would go a long way for good eel— we worked out the skeleton of the methods. A "to do" list was drafted (during the next few weeks, it would be redrafted many times, but looking back on it now, that first draft was pretty accurate). We worked out how the Celera crew would be trained to annotate *Drosophila* sequences, distinguish the genes from the crap, recognize where the gene prediction programs join two genes together or split one apart, and recognize the real EST hits from the false ones. Trained, and then tested. We worked out how we would divide these predicted genes among the experts whom we planned to invite to the jamboree. This was a strategy never used before, totally untried and untested.

Worse, it meant that I had to get back home, do some serious work, and stop jetting around trying to get other people to do the real work. It all relied on a new database that Suzi and I were trying to build, in collaboration with colleagues who work

---

[92] The DNA Data Bank of Japan is at the National Institute of Genetics (NIG) in Mishima. A very nice small town and the NIG has a wonderful collection of cherry trees. The director of the NIG was Yoshi Hotta, yet another drosophilist from the Caltech mafia. Graham Cameron was then my colleague at the EBI, where we were "joint-heads." That title was not of our choosing.

on mice and yeast.[93] It is called the Gene Ontology (GO) database,[94] and I am afraid that the name was a deliberate (and successful) attempt to annoy people who will remain nameless (they know who they are). It's really very simple—just a standard set of concepts and the relationships between them, so that we can describe the function of gene products.[95] I had a month to do this A long conference call the next day (I was back in Cambridge but Suzi had stayed on at Celera) gave us the first draft of the allocation of predicted genes to groups of GO terms. Sima had joined Suzi from Berkeley, together with her sister to look after the young Aden. Sima and Jennifer[96] had independently annotated some Berkeley sequence and were there to compare results. The jamboree was now set for the second and third weeks of November. That presented both Gerry and I with a problem, because we were both committed to be in London in the middle of this period. Gerry immediately started a campaign to get out of this—he was going to speak at a conference that I had helped to organize.[97] "You can speak for me," he said. Yes, I could, and would. But dammit, I was not going to give in that easily. I then started my own campaign: Would Craig fly me on the Concorde back for the day? I knew that was silly, but I

---

[93] Particularly David Botstein and Mike Cherry of the *Saccharomyces* Genome Database in Stanford (yeast) and Janan Eppig and Judy Blake of Mouse Genome Informatics at The Jackson Laboratory in Bar Harbor, Maine.

[94] http://www.geneontology.org/

[95] At the time, the Gene Ontology consortium was unfunded, except for a relatively small, but very useful, gift from AstraZeneca (a big pharma), courtesy of Ken Fassman. In October, Gerry wrote to Craig suggesting that Celera might like to donate $100,000 to us during the course of 2 years; as far as I am aware, there was no reply. Nevertheless, without the work we did on the GO, the annotation jamboree would have been chaotic.

[96] Jennifer Russo Wortman was a senior member of the Celera chromosome team and, later, my "buddy" (see footnote 140).

[97] With Matthew Freeman of the Laboratory of Molecular Biology in Cambridge, this was a 1-day meeting on *Drosophila* for the Genetical Society. I presented Eric Weischaus with the Society's Mendel Medal.

thought that he would settle on business, rather than consigning me to 42H again. He did.

In early October, Suzi and I needed to be in Bar Harbor, Maine to meet with our colleagues to discuss the Gene Ontology project. OK, we will meet in Boston and drive up, and that way we can at least have 4 hours to work together in the car each way. I had not reckoned on Suzi having friends in rural Maine, nor on the fact that if you walk across fields in Maine in the fall you get attacked by chiggers.[98] Now, you may not know that chiggers are mites. To a drosophilist, mites are what crabs are to a sixteen-year-old boy. Bad news. For a drosophilist to have mites burrowing under his skin, laying eggs, is a really disturbing experience. Thank you, Suzi, for that. But we did work—with me driving and Suzi taking notes on her palm pilot—there and back (with a culture stop to see the Wyeths[99] in Rockland and a pit stop in Bath, Maine to have oysters. Unfortunately, they were out of season). We stopped for a quick visit with Bill in his new house in Wayland to brief him and then I ambled down the Mass Pike to Logan, only to look at my ticket while driving through the Callaghan Tunnel[100] to find that my British Airways (BA) flight left at 8:10 p.m., not 9:10 as I had thought. It was then 7:20. BA rose to the occasion by upgrading me. I itched all the way to Heathrow.

A week later, MarkY and Suzi are in Hinxton. More planning, much of it over a curry in the Curry Queen.[101] What are we going to do about protein domains? Many people have made databases of protein domains: little bits of sequence that

---

[98] That is what I thought they were at the time; in fact, they were female *Simulium*, otherwise known as blackfly.

[99] Heroes of Maine. A family of painters. Some are OK, others are awful, but the gallery is great, in a large clapboard church. Rockland is worth a visit. It has not been tarted up and has an extraordinary Main Street, which, for once, is not Maine Street.

[100] Actually, it was a Sunday and I could have used the new Williams Tunnel, but I did not.

[101] On Mill Road in Cambridge. One of the better ones, if you avoid the vindaloo.

give one an idea of the structure, and even function, of proteins, the ultimate products of genes (I simplify but, frankly, I do not have patience to write a text on molecular biology). With names like PROSITE, PRINTS, PFAM, BLOCKS, PRODOM, and so on, they are the Tower of Babel of the protein database world. But Rolf[102] had this great idea: Let's all get together and make one database of protein domains. Let us share and integrate (integrate is a *really* sexy word in bioinformatics; the funders love it). So INTERPRO was born. INTERPRO was about to go to beta release (that is a release that might work, but no guarantees. In contrast, nobody in their right mind would touch an alpha release). It's all so obvious, so I drag Suzi and MarkY to the EBI for a demo of INTERPRO. We meet Wolfgang[103] (for some reason, Suzi always calls him Wolfi), who gives a live demo of INTERPRO. "Mark, we must have this. Phone MarkA and tell him to send a ticket for Wolfgang; he can install the code at Celera." MarkY, MarkY who cut his bioinformatics teeth at Washington University,[104] is impressed. "Yep." So we get INTERPRO. "But we must have Ewan, says Suzi." Ewan[105] is a hidden Markov model expert from next-door in the Sanger Centre, Ewan is the keeper of PFAM, Ewan wrote GeneWise. Ewan is hot property, and the kid does not yet have his Ph.D. There is some sensitivity here. Ewan works for Richard,[106] who works for John, who works for

---

[102] Rolf Apweiler, head of the sequence database teams at the European Bioinformatics Institute (EBI).

[103] Wolfgang Fleischmann worked for Rolf at the EBI.

[104] Washington University, St. Louis, was then home of Bob Waterston and a major sequencing lab. Bob's and John Sulston's groups sequenced the 100-mb worm genome (see *Science*, December 11, 1998). Gerry said we, i.e., *Drosophila*, have to do better than that; we did. Bob and John wrote a letter to Jim Watson at the NIH saying that no money should be given to fly sequencing (i.e., it should all go to the worm). Don't deny it, guys—I have a copy.

[105] Ewan Birney now works at the EBI.

[106] Richard Durbin, head of informatics (John Sulston's deputy) at the Sanger Centre.

Ewan Birney

the Wellcome Trust, whose relationship with C
whole Celera deal is somewhat cool[107] (I really mu
English habit of understatement). We get Wolfgan
although MarkA is heard muttering about his budget. Budget?
Fudge-it.

Next stop is Zurich to meet Gerry. This was now getting seri-
ous. We had to find 40 or so colleagues willing and able to
spend a week or two at Celera in 4 weeks to work on the
sequence. A letter of invitation had to be drafted, sent to
Celera for approval, and sent out. What experts did we need?
One for transcription factors? Of course. And what about
immunity? Young Bruno[108] is here from Paris: Shall I ask him?
(He phones his wife, and she says yes.) Oh, and we must have
George, who knows about human disease genes, and
Michael[109] for DNA replication. And so the list was made up
and we tapped out a draft invitation letter on my G3[110] sitting
on some steps in a hall, young drosophilists of Europe milling
about us, astounded that these old men could actually use a
computer. Gerry sent the draft to Celera and spent a few days
on the phone priming the invitees. It was looking good—everyone was very enthusiastic and nobody turned us down except
for a few who had unbreakable commitments to teach (mind
you, we had written "The opportunity of a lifetime" as the sub-
ject line of this e-mail). Celera approved the letter, including the
promise that all of the annotation data would be made freely
available without let or restraint. They agreed to pay everyone's
expenses (but no consultancy fees, which was fine, indeed very

---

[107] See *The Common Thread*, footnote 1.

[108] Bruno Lemaitre, from the Centre de Génétique Moléculaire in Gif-sur-Yvette, southwest of Paris. Spent a summer as an undergraduate student in my lab. The CNRS (Centre National de la Recherche Scientifique) has quite a good canteen at Gif, but do not plan to do any serious work after lunch.

[109] George Miklos was at the Australian National University in Canberra and then with Edelman in La Jolla. Now claims to have retired, but has a con-sultancy company in Sydney. Mike Botchan is from Berkeley.

[110] A Macintosh laptop computer. Mine was the heavy old black one.

good—we were not being bought). Craig added the word "substantial" to our "contribution" as the buy-in for being an author on one of the papers.[111] We agreed that everyone would sign a simple nondisclosure agreement, undertaking not to disclose any data before its publication. For the first time, I believed that it would really happen.

But it nearly did not. My life was getting out of control. The annotation was to begin on Sunday, November 7. I was leaving Cambridge on Wednesday, November 3 for a meeting in Bloomington.[112] On Thursday, November 4, *Nature* was running an editorial and two news stories on the fiasco of the failure of the European Commission to fund the infrastructure for biological research in Europe.[113] The previous Thursday, someone[114] had the bright idea that my boss, the Director-General of the EMBL,[115] should organize an open letter for publication, signed by the Great and the Good. "You do it, Michael. I do not have time." It is not often that I swear at a director-general. *Nature* goes to bed at 4 p.m. on Monday afternoon. At 2:30 p.m., I was still phoning the Great and the Good all around Europe. I was also trying to get my act together so as to spend more than 2 weeks in the U.S., write a talk for Bloomington, and finish off some preparatory work for the jamboree.

Disaster strikes at about 3 p.m. Alan[116] e-mails me to ask if I

---

[111] There had been something of a feeding frenzy between *Nature* and *Science* for the privilege of publishing this work. Mark Patterson, an editor at *Nature*, thought that he could call on the "old boys" act and persuade me to convince Craig to publish it in *Nature*. I tried to rid him of this delusion, but I think he still believes that I let the side down. In fact, Craig had already made a deal with *Science*.

[112] The meeting was to celebrate John Law, an insect physiologist in Tucson, Arizona. The Director-General (see footnote 44) was there with his wife Sarah, also guests of Peter and Lucy (see footnote 121).

[113] See Editorial 1999. Vacuum at the heart of Europe. *Nature* 402: 1.

[114] Declan Butler, European correspondent for *Nature*, on the lookout for a good story, as always.

[115] Fotis Kafatos, see footnote 44.

[116] Alan Robinson was on the staff of the EBI.

had seen the new *Drosophila* Web page on the NCBI's[117] server in Bethesda. No, I had not. "The fucking bastards—they have screwed us": There, for the world to see, was 11 million base pairs of Celera's *Drosophila* sequence, available only via a special server and covered with protective notices: "THE SEQUENCE INFORMATION PROVIDED IS FOR YOUR PERSONAL AND NON-COMMERCIAL USE. YOU MAY NOT COPY, DISTRIBUTE, TRANSMIT, REPRODUCE, SELL...ANY OF THIS SEQUENCE INFORMATION, IN WHOLE OR IN PART." Not freely available. Not an A, G, C, or T. Just what I had, last May, told Francis that they would do and what I had hoped that he would stop, for the NCBI is, after all, an agency of the NIH. I react badly to this sort of stress. I screamed for Gerry. But Gerry was nowhere to be found. He was later discovered in Philadelphia, of all places. Bill was upset (again, I understate). I phoned everyone I could. David[118] would not take my calls. I spoke to Jim[119] (Jim2; he works for David at the NCBI), who said, "But I thought that this had the support of the fly community." I will suppress my reply. I phoned Francis. Francis was in conference (Francis is always in conference). I told his aide that it was life or death. (I never could understand that TV insurance company advert that promised "We never make a drama out of a crisis"—why not?). Francis sent me an e-mail. Gerry was tracked down. "Don't panic, keep your head, stay cool." "But they have screwed us. I am going to e-mail all invitees to the jamboree telling them not to come. It's off, over, kaput, finished, done for, dead as a parrot, a non-jamboree." "Keep calm, Michael. Let me deal with this. I will sort it out, promise." He did, of course. Within 36 hours, the data were where they belonged, in GenBank, free to the world, and the offending

---

[117] The National Center for Biotechnology Information is the "EBI of the U.S.A."

[118] David Lipman, director of the NCBI. Graham's and my equivalent.

[119] Jim Ostell, who, under a different name, was a graduate student of Fotis, when Fotis was a professor at Harvard. Like Francis, he is tall.

(and offensive) NCBI page had been pulled. I trashed the e-mail I had prepared and went to Bloomington, leaving Graham to talk to the press when *Nature* hit the streets. And Celera? Why had they done this? Was it just a misunderstanding? Probably. After all, they were also under enormous pressure and even Craig and MarkA must delegate. The alternative is that Craig was just trying to get my goat, but you would not do that, would you, Craig? Of course not. One thing international yacht racing teaches you is respect for the rules of the game.[120]

Only one more hiccup to go. Peter and Lucy,[121] my dear hosts in Bloomington, let me check my e-mail from their computer. A long e-mail from Eula,[122] MarkA's personal assistant. In some god-awful Microsoft program, Word98 most likely (we all hate Microsoft). I cannot decode it. Phone Suzi: What does it say? "It is the nondisclosure agreement that we have to sign." "Can you read it to me?" It's full of parts and parties—"this party on their part"; prison if you reveal where the lavatories are in the Celera building. Not only can we not disclose confidential information but, also, such confidential information is provided "AS IS" and comes without "warranties, express, implied or otherwise." "But MarkA said that it would be simple, not this lawyer crap. And where is Delaware, anyway? I don't mind; I would sell my mother by now, but the kids are going to be scared silly at being told to sign this." OK, I have learned my lesson now: Tell Gerry. Don't phone MarkA and harangue him, just tell Gerry. Gerry is at Cold Spring Harbor (he loves that

---

[120] Yacht racing has a complex set of internationally accepted rules, as I learned in my youth, much of which was spent messing about racing boats. I should add (in August 2005) that Jim Shreeve spilled the beans in his book (see footnote 1)—Craig was indeed trying to see just what he could get away with.

[121] Lucy and Peter Cherbas. Peter was a graduate student at Harvard and my postdoc. Lucy was a postdoc with Richard Jackson in biochemistry at Cambridge.

[122] Eula Wilturner. She was terrific when the jamboree took place.

place, having been an URP[123] there nearly 30 years ago). Gerry is taking a bioinformatics course, learning the difference between BLASTP and TBLASTX, and what Z scores mean. Good for him. Gerry is losing patience; I can tell from the word spacing in his e-mail: "Don't panic, it's only boilerplate, no one will worry about it." They didn't, with one minor exception (not that the excepting person was minor,[124] just that his exception was minor). OK. I relax and go out to throw Frisbees for Peter and Lucy's new dog.[125]

Behind the scenes, real work was being done. On the Thursday before the jamboree, Martin Reese and David Kulp at Neomorphic got word from MarkY that the assembled *Drosophila* sequence could be downloaded from the Celera FTP site. Now came the real test of Genie, their gene prediction software that we had assessed in the GASP meeting. It almost proved too much for Neo's Linux cluster, and there was no time for mistakes and little time for tweeking the software. The computes did not end until the morning of Sunday, November 7, just in time for Martin to catch a flight to Dulles from San Francisco, where he immediately had to parse Genie's ouput into a format that the Celera software could digest. Meanwhile, MarkY, Suzi, and Peter Li had a terrible job hacking all of the files sent by the outside attendees, files of their favorite proteins that we needed for the analysis. Some of these (horror of horrors) were in Excel spreadsheets, a computer scientist's least-favored format. MarkY also ran Genscan, another gene prediction program, but trained on a human gene set. In the event, Genie did well, predicting 13,601 protein coding genes; Genscan predicted over 19,000. When 13,601 turned out to be right on the ball, Martin was beaming.

---

[123] A participant in Cold Spring Harbor Laboratory's Undergraduate Research Program.

[124] Pat O'Farrell, from the University of California, San Francisco.

[125] I mean the plural. You must have two Frisbees to tire out the dog, called Arge, a black lab. His namesake is a famous Greek molecular biologist who always dresses in black. He (the famous Greek) is now at Columbia, but was with Fotis at Harvard. A previous dog had been called Wally, which caused great

Sunday, November 7—the day the jamboree is to start. Was I nice earlier about United Airlines? Well that day they returned the favor. Five and a half hours on a Sunday afternoon in Indianapolis airport.[126] And just to get to Dulles. I even have to admit to going to a Borders. A nice Afghani driver from the Washington Flyer (how do they come to have a monopoly on cabs at Dulles?[127]) gets me to the Marriott Washingtonian in Gaithersburg by midnight. I'm all bright-eyed and bushy-tailed at Celera by 8 a.m. (thank you, MarkY, for the pickup, but why do you drive a truck with only two seats?[128]). The jamboree has started. It is, of course, absolute chaos, but at least it is the sort of chaos that with a bit of work can be brought to a semblance of order. And we have 2 weeks to bring this off.

"If you want anything, just whistle."

"You know how to whistle Michael, don't you? Just pucker your lips and blow."[129]

---

confusion when called to heel in the courtyard of the BioLabs at Harvard.

[126] Indianapolis Airport is not as bad as Detroit's, but it is pretty bleak. Avoid it, if possible (that is, drive to Chicago).

[127] There is a story here. The Washington Flyer was then owned by an Afghani, I believe, and all the drivers are from that general neck of the woods. Talking to them, you learn a lot about the politics of the region. On a later trip, I had a very nice Pakistani who was very illuminating, though rather depressing, regarding the recent Pakistani coup. I learned much later that Jon Snow, now an anchor at Channel 4 News in London, also valued conversations with the taxi drivers in Washington (see Snow J. 2004. *Shooting history*. Harper Collins, London).

[128] The problem, I should perhaps explain, is that there were three of us. With gallantry, I allowed Suzi to sit on my lap. This seemed to provoke undue hilarity among the commuters of Gaithersburg and we nearly caused an accident crossing Shady Grove. Shady Grove achieved mythical status during the jamboree, probably because Martin Reese kept getting lost in it. This lead to the instruction to all attendees to "never turn on Shady Grove."

[129] I trust that this quotation needs no reference. OK, then, for pedants and the ignorant, it is Miss Bacall to Mr. Bogart in "To Have and Have Not" (1944): "You know, you don't have to act with me, Steve. You don't have to say anything and you don't have to do anything...not a thing. Oh, maybe just whistle. You know how to whistle, don't you, Steve? You just put your lips together and blow."

Photo Gallery

Participants of the 1998 Cold Spring Harbor Laboratory Genome meeting relax on Blackford lawn overlooking the harbor.

Orthodox Academy of Crete at Kolymbari (*right*), Greece with Ghonia Monastery to the left. The low building in the middle is the Academy's original conference center and the larger building to the right is the new extension.

Photograph by Jane Bown (©1988) of six Cambridge drosophilists in the Medical Research Council Laboratory for Molecular Biology in 1987. (*Left to right*) Michael Akam, Michael Wilcox, Alfonso Martinez-Arias, Michael Ashburner, Michael Bate, and Peter Lawrence. With the exception of Michael Wilcox, who died in February 1992, all remain at Cambridge. (Reprinted, with permission, from the February 7, 1988 London *Observer*.)

(*Left to right*) Mark Adams, Peter Li, Michael Ashburner (*seated*), and Gerry Rubin in the Geneva conference room of Celera during the jamboree. (Photo by Ken Burtis.)

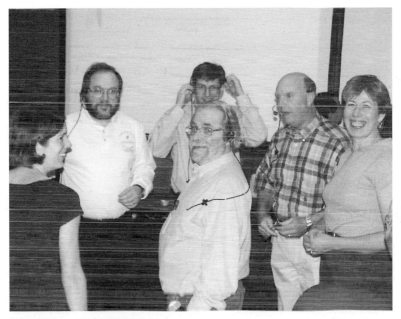

Participants of the Celera jamboree, November 1999, in the "control room," where it was initially planned that a room full of operators would be available to answer queries related to genome analysis. (*Left to right*) Leslie Vosshall, Gerry Rubin, Mark Adams, Michael Ashburner, Craig Venter, and Suzanna Lewis (Photo by Ken Burtis.)

Genome Project members Susan Celnicker, Gerry Rubin, Mark Adams, Gene Myers, Suzanna Lewis, and Craig Venter after publication of the sequence. Note the CD of the sequence in Craig's hands. Pittsburgh, March 2000. (Reprinted, with permission, from *The New York Times*, April 11, 2000 [Gary Tramontina/*The New York Times*].)

3

Puffing and Panting: The Jamboree

First things first. Where is the expresso machine? We are on the third floor of Celera's building 2—almost an entire floor divided into cubicles (henceforth, cubes)—pure Dilbertville.[130] Not bad, a coffee machine in the kitchen, but not serious coffee. I ask around and find an expresso machine on the fourth floor. It is only several days later that we find this to be the personal property of (most appropriately) Celera's in-house patent attorney, Richard Millman. But he is a nice guy and allows us to use his machine (and his beans). A good start.

Let's give credit where credit is due. Celera really pulled out the stops for the jamboree. We each have our own cube, a phone, and a computer. Everything possible is done to make us feel at home and allow us to work productively. The politics are over. Now we have fun. There must be 60 of us all together—20 or so Celera staff under Peter Li and nearly 40 others, mostly (but not all) fly people. Jennifer and MarkA from Celera run the computes, Peter Li writes some nice software so that we can see what the computers tell us (for each predicted gene there is a "jam page," an HTML page that summarizes all of the data about it), and Gregg[131] has come from Neomorphic to tweak

---

[130] See www.dilbert.com
[131] Gregg Helt had been a graduate student of Gerry's and, with Martin Reese and Cyrus Harmon, founded Neomorphic Software, Inc.

PETER LI PORTRAIT   OCT 30 CSHL (DAVENPORT).   LEWIS MILLER '05.

Peter Li

the browser. Suzi is den mother and coder extraordinaire. The biologists break into two groups: Most of the Celerites check gene models. These gene models are then used to predict the proteins coded by each gene. MarkY had written a program (for reasons I never discovered, he called it LoveAtFirstSight), which then grouped these proteins into bins based on their predicted Gene Ontology classification. Most of the other group, the visitors, spend their time going through the bins of genes that are of interest to them—transcription factors, proteases, cytochrome P450s, etc. Within a day or so, a routine develops. A morning on the computer and then (at least for a few of us) a quick trip to a sushi bar for lunch. More analyses until about 4 p.m. when Peter Li sounds his large Chinese gong as a signal to assemble in Geneva (for some reason, all of the meeting rooms are named after cities) for the daily bull session[132] to consider where we are that day. What are the problems? What computes should we do? Craig makes an appearance most days. One day, we are invited to see a visiting wolf. "His name is Francis," says Craig. Was this a mate for one of Craig's famous poodles or just a warning that we should keep in line? Back to work. Dinner is usually take-out. We take great advantage of the ethnic diversity of the group, with great and gargantuan Chinese, Indian, and Italian meals. Back to work and, with luck, back to the Marriott by 2 a.m. Yes, that is a downside—even for a Marriott, the Washingtonian in Gaithersburg leaves much to be desired. At least I have a smoking companion in the shape of David Coates.[133]

It's just like being a postdoc again. Wonderful. Apart from the daily call back to Cambridge to make sure that disaster has

---

[132] "Too many bulls, not enough cows," MarkA said after one particularly tough bull session.

[133] A protease expert from the University of Leeds. Later, to my disgust, became a dean, ruining his research life.

PORTRAIT OF MARK YANDELL   NOV 3 (CSHL)   LEWIS MILLER '05

Mark Yandell

not struck, we are all wholly dedicated to the job. Gene number, of course, is a big issue. Gerry, Bill, and I have a bet on it, witnessed by Thom.[134] The winner gets dinner for two at the restaurant of his choice.[135] Gerry has bet low: greater than 10,000 and less than 15,000; Bill has bet high: greater than 20,000. I am in the middle. But coverage is also important. Several thousand fly genes have been sequenced by the community in the last 20 years or so. Do we have them all? I get the data together and Mike Cherry[136] runs the compute. At an afternoon session in Geneva, we see the data and only very few genes (19 at first count) are missing. My notes of that day's bull session include "Which is amazing!" We weed out those that are really from the chicken and then soon find all but two: George Miklos and I immediately identify these as fake. Yes, I am afraid that even the fly community has its fraudsters.

Gerry arrives late to the jamboree. Soon takes me aside and swears me to utter secrecy: He has accepted the post of vice president for biomedical research with the Howard Hughes Medical Institute[137]; he will go on half-time leave from Berkeley. I try to tell him the fallacy of two half-time jobs, but he is raring to go. The news is public a few days later and Bill sulks because he was not in on the secret. Grown men.

But we paid the price for such a rushed planning period. No, that is not altogether fair. Nothing like this had ever been attempted before and we had no guidelines. We had sent Sima out to Celera to give a week's annotation training in October. One problem was with the software. And, as usual, software problems only become apparent when the software is used in anger. Gregg, who has mastered the art of no sleep for days on

---

[134] Thom Kaufman, evodevo genius in Bloomington and a colleague on FlyBase.

[135] I am open to suggestions.

[136] Mike runs the yeast genome database (SGD) at Stanford. Unflappable.

[137] A major biomedical charity in the U.S. Founded with the dollars left by the legendary Howard Hughes.

end, releases new versions of the browser on an almost daily basis. The jam-pages, which we used to view, edit, and store our analyses, needed a lot of work. We need immediate updates. We are getting them every night; we compromise on hourly updates. Peter pulls a face. Frankly, it was only in the last couple of days of the jamboree that we really had the computational tools that we needed. But we make do. Another problem is with annotation quality—we appoint Takis to be quality manager for gene models. He comes up with guidelines and "six steps for SAGS."[138] Many of us biologists learn a useful lesson: Those who can really make the most interesting discoveries in data of this sort are the ones who combine biological insight with an ability to hack Perl. Ewan does a nice collaboration with Leslie Vosshall[139] and writes an HMM[140] for olfactory receptors—yes, flies can smell. He gets a bit carried away and wants to run his pet software GeneWise on all the genome. The problem is time; GeneWise is very compute-intense. But that does not phase Ewan (not much does): "If Celera does not have the juice, I know some supercomputers at Cambridge that definitely have the power to chew through this...if it is not done for the paper then I propose to do it once the genome is out and [*grin*] embarrass you all [*grin*] by finding more genes." That's enough to give you a flavor of Ewan's style. Rolf and Wolfi do a wonderful job mapping protein domains from INTERPRO to all of the predicted genes; only took 45 hours to compute. I seem to spend all day on screeds of Microsoft Excel spreadsheets of data sent to me by Jennifer.[141]

Controlled stress release is very important during intense periods of research like this. Celera, of course, had planned for

---

[138] Successfully annotated genes. Very Takis.
[139] Leslie is at The Rockefeller University in New York and an expert on the sense of smell in flies.
[140] Hidden Markov model. Don't worry about it.
[141] Each visitor to Celera had an inhouse "buddy." Jennifer was mine, which was great, both scientifically and because she had a very sympathetic shoul-

this. Scattered around the tops of the cubes are plastic guns for war games on the lawn outside—or when things got really tough, in the lanes between the cubes. We party, taking over a restaurant in Gaithersburg and Mark Adams' house. We have a photo op in Mission Control, Celera's high-tech control center. Lots of fooling around: I persuade Craig and Gerry to wear headsets, pretending that they are Gene Kranz[142] and friends.[143] We freak out over uni at lunch. One couple even became engaged to be married.[144] (How did they find the time or energy?)

But we make it...almost. It is pretty clear by the end of the fortnight that there is still a hell of a lot to do. Some things never got done. I had planned to do an analysis of the repeated sequences in the genome. We invited Tom Eickbush[145] from Rochester to look at the transposable elements. Unfortunately, the precomputes were done by a Celera scientist who spoke no English. Thom left in disgust after 2 days and I gave up on the idea. Suzi and I agree to come back in early December for a planning session (just what do we need to do before we can write this up for publication?). I had to be in Washington D C at that time anyway for a meeting on the malaria genome (yep, you guessed it, another Marriott). Suzi flies in and we spend a day at Celera with Peter, MarkY, and Jennifer, taking stock. But we are in for an unexpected surprise. We are invited to the Celera Christmas party at the Sequoia on K Street. Supping

---

der onto which I could weep when it all got a bit much (about twice a day).

[142] Flight director for the Apollo 11 mission; see Kranz G. 2000. *Failure is not an option: Mission control from Mercury to Apollo 13 and beyond.* Simon & Schuster, New York.

[143] For the photos, see Pennisi E. 2000. Ideas fly at gene-finding jamboree. *Science* 287: 2182–2184. The photograph of Craig standing on my shoulders has never, unfortunately, been released, and is said to be "lost." It would, of course, have been captioned "Standing on the shoulders of a giant" with apologies to the late Robert Merton, whose book *On the shoulders of giants: A Shandean postscript* (1965. Free Press, New York) is a great read.

[144] Mark Fortini and Debbie Morrison.

[145] Tom is at the University of Rochester and an expert on the evolution of transposable elements.

JENNIFER RUSSO WORTMAN   CSHL 31 OCT (DAVENPORT)   LEWIS MILLER '05

Jennifer Wortman

with the Devil? You bet. Suzi, of course, has her glad rags. I reckon that playing the English eccentric card will make up for my lack of sartorial elegance. The atmosphere is pretty wild. Celera stock jumped nine points the day before and everyone had the ticker display on their screens. Craig admits that even he is now taking the analogy with Bill Gates more seriously than when the press first suggested it. Now that is quite a scary thought. Moreover, there had been much knocking of Celera on December 2 when the Human Chromosome 22 paper was published in *Nature*. Even Francesca[146] calls me a class traitor when we speak that morning. Strong stuff, but none of *my* family were party members.

But we are still exhausted from the jamboree and the relentless travel. Suzi and I pay a visit to Jim2 and David Lipman to discuss how we can get the data into GenBank for public release. David to Jim2: "Have you told Suzi the news?" Jim2 to Suzi: "Incyte has just deposited [in GenBank] the entire sequence of *Drosophila*." Suzi is so out of it that she believes him, for a minute or two.[147] That cost Jim2 a vodka and tonic. But back to reality, we have much to get done. The Celera team (the "chromosome team") is told that Christmas has been cancelled for 1999—1 day off, that is all. I am as dedicated to science as the next man, but I am buggered if I am going to do that. Off to Ireland for the New Year and a family wedding. I almost don't make it—the strain is beginning to take its toll and I come down with pneumonia during Christmas. "If I knew I was going to live so long, I would have looked after myself better."[148]

---

[146] My wife.

[147] In Suzi's defense, I think that she had not slept at all for 36 hours.

[148] Attributed to Micky Mantle.

couple of days in a guest house in Vienna, Maryland that stank of cat urine and had the toilet in the corner of the room behind a screen and watching the whistling swans[151] and Canada geese[152] on the banks of the Nanticoke River.

It was clear that more work had to be done on the gene models. They were simply not good enough to give us reliable proteins. Ideally, we would have liked to repeat the jamboree in Gaithersburg, but that was pushing the budget (and time) too far. Celera's parent company, Perkin-Elmer, had a place[153] at Foster City in the Bay Area. It was agreed that they would run a short jamboree 2, with Ken Burtis[154] providing computer access for the curators at UC Davis, lead by Sima Misra—without question, the queen of *Drosophila* gene annotation.

One last visit to Celera at the end of January. It was awfully inconvenient for Jennifer, because she had family plans and "canceling the trip entirely is of course possible. It would just make me persona non grata with the folks [her parents]." I felt awful, but we were simply too stressed out about the data quality question and needed to meet. We had to double-check all of the computed translations of the genes, translations that absolutely depended on the correctness of the gene models, and we needed to check all of the links between the gene models and FlyBase. It was dreadfully tedious, but the trip was made more interesting by a spectacular ice storm on Sunday,

---

[151] *Olor columbianus*, probably conspecific with the European Bewick's swan.

[152] *Branta canadensis*, probably the small interior subspecies. See http://www.sibleyguides.com/canada_cackling.htm

[153] Applied Biosystems, providers of the sequencing machines.

[154] Ken is a *Drosophila* molecular biologist, now a dean at University of California, Davis. As a postdoc with Dave Hogness at Stanford, he cloned the early ecdysone responsive genes from *Drosophila*. I had spent the best years of my life working on these and had proposed that they encode transcription factors. Ken's early data almost shortened my life because he could find no evidence that they were protein coding. Luckily, he soon discovered what he had done wrong.

January 30.[155] At about 3:00 in the afternoon, someone hiked to the nearest window and suggested that those who had to drive might consider doing so now. She was right—it took Suzi and I over an hour to get back to the Marriott. The next morning, only MarkY's truck saved us from a wasted day; it also took us to a last, but memorable, Gaithersburg sushi meal, at which we were offered and accepted, I am ashamed to say, live lobster. Still, it gave me a chance to show off my knowledge of crustacean anatomy (and was, I admit, delicious).

After this trip, the panic level, for me at least, slowly declines. True, there are some papers to write. Craig and Gerry have arranged for them to be published in *Science* to coincide with the *Drosophila* Research Conference, this year to be held in Pittsburgh. We all write little bits (MarkA calls them "vignettes") to be stitched together by Gerry and MarkA, but I am hardly involved. Catherine Nelson[156] is employed as a consultant to put all of the bits together. So I get e-mails like this: "Thank you for your paragraph about clusters. I think it's an interesting topic and I want to give it as much space as possible, but..." Catherine: Thems was my words, thems was. Keep cool. There is, of course, horse trading over authorship and author order. I come under some pressure (again, for American readers, I understate) from my colleagues on the European program to see that they get some of the glory. True, we did provide some sequence from the X chromosome and lots of BAC ends, but they were not actually used in Gene's assembly. But the spirit was there and eventually we win, simply, I think, because everyone else involved is too exhausted to argue the toss. We also think that there should be a paper from the European group and we cobble one together. Most of that work was done by Melanie Gatt[157] and Takis Benos.

---

[155] See http://www.kiat.net/dc/images/weather/snowtotal013000.gif

[156] Catherine was then a postdoc in Suzi's group with Gerry.

[157] A postdoc with David Glover in Cambridge. A New Zealander with a penchant for belly dancing.

All I did was to provide a title and negotiate publication with Barbara Jasny, *Science's* editor for genome papers.[158]

Writing papers was only half the story. These papers could only possibly give the highlights; the *real* information could only be presented through a database, FlyBase. It had been agreed that Celera would, on publication, deliver all of the sequence and the annotation to FlyBase. Publication had been synchronized with the annual *Drosophila* Research Conference in Pittsburgh and we knew that our colleagues attending would be hungry for the detailed data as soon as they arrived. It would be very embarrassing if this was not done; the conspiracy theorists would have a field day. Sitting in the back of a Washington cab, Suzi and I had sketched out a design for a Web interface for the data and Suzi had asked Chris Mungall[159] to implement this. Two days before the Pittsburgh unveiling, MarkY made good on Celera's promise for the data. While Gerry and I were locked in with the FlyBase advisors, Suzi was constantly on her cell phone to Chris back in Berkeley getting status reports. It was touch and go, but Chris pulled it off beautifully and the new Gadfly database went live at Berkeley right on schedule.

It seems too predictable to tell the W.C. Fields joke about Pittsburgh. But he was quite correct.[160] It does not even have a direct BA flight from Heathrow. I have blanked out about whether the hotel was yet another Marriott. It was grim. Fly meetings are always a strain and this was more so than most. The

---

[158] Barbara and I were to have somewhat of a falling-out in 2001 over the matter of the publication of an advertisement from Celera masquerading as a scientific paper on the human genome. I tried to arrange a boycott of *Science*. It did not work, but it is interesting that *Nature* has now (2005) published about twice as many "whole-genome" papers as *Science*.

[159] A brilliant young computer scientist from Edinburgh working for Suzi, distinguished, then, by his purple hair. Chris had once worked in the office next to Ian Wilmut, creator of Dolly the lamb.

[160] For many years, I believed this story. But I am now convinced that in this apocrypha, Fields referred to Philadelphia, his birth place. Pity.

publication of the fly genome occurred on the Friday. Late Thursday night, Craig and the suits jetted in, flying private, of course. The Celera publicity team is *really* good; one has to admit that. There were CDs of the genome on every seat and lots of toys, including a jigsaw-puzzle version of one of the big figures from the *Science* paper. Lots of razzamatazz. I was ill, retired to my room to sulk, and caught a dawn flight to J.F.K. the next day hoping to catch the day plane home. That was full, so I spent the whole day in the BA lounge waiting for the first evening flight at 6:30 p.m. It was done. Not bad timing. Back in October, we had scheduled publication for February 15, just 5 weeks late. But I was also done in. For the next 18 months, Gene Peykal, professor of clinical psychiatry at Cambridge, tried to mend my head, until the penny dropped and my general practitioner found that I was walking around with a blood glucose concentration of 35 mM.

POSTSCRIPT

This a story about people, particularly about how people act and interact at times that are, for them, full of great excitement and stress. But this is also a story about science and of the social interactions that make science fun. It is the importance of these interactions that makes scientists such wonderful gossips, about science and people. No one should walk away after reading this book with the belief that all science is like this. Much of it is still a very lonely activity. In my field, however, international collaborations, involving many people, often with competing objectives and interests, are increasingly becoming the norm. Make a public call for a genomicist on the concourse of terminal 4 at Heathrow or terminal 7 at J.F.K. any day, and many would come forward. Many of these collaborations are on a knife-edge—desperately trying to remain cohesive to both get the job done and give the funders confidence, against the strong centrifugal pull of competitive egos and conflicts of interests.

This story ends with the publication of the *Drosophila* genome on March 24, 2000. But the world has moved on since then. The science has moved on. The sequence of this fly pub-

...s been revised, errors have been corrected, and
filled. Its annotation has been completely over-
...me not by a jamboree, but by the dedicated
effor... ...ut ten FlyBase curators in Berkeley, Harvard, and
Cambridge, lead by Sima Misra, over a 12-month period
(and published in December 2002 in *Genome Biology*). The word
"jamboree" has entered the common vocabulary of the field,
and many other annotation jamborees, for the genomes of
other organisms, have been held. The companies and institutes
have moved on. Celera is now, like Applied Biosystems, a division of Applera, but it is no longer a genomics science-driven
company; its mission is to "discover and develop meaningful
new therapies that improve human health." I am sure what is
done there is very worthy, but it is no longer at the cutting edge
of science. As will be seen below, most of the Celera players in
this story have left the company for academia. As an ironic footnote, I joined the Science Advisory Board of Applied Biosystems
in 2005. The Sanger Centre is now the Sanger Institute under a
new director[161] and has greatly expanded in size. The Berkeley
Drosophila Genome Project closed its doors (and network ports)
in November 2005. Neomorphic Software, Inc. was purchased
for an absurd sum by Affymetrix in September 2000. And the
people have moved on. Here, in brief, is where they were then
and where they are now, in August 2005, or—for a few—where
they will be by the time this is published.

Mark Adams left Celera in 2003 to become a professor in
the department of genetics at Case Western Reserve University
in Cleveland, Ohio. There he works on mouse models of human
disease. **Rolf Apweiler** is still at the EBI doing a great job running all of the protein and nucleic acid sequence data banks.
Michael Ashburner: After October 2001, I returned full time to
my position in the University of Cambridge, after six years of

---

[161] Alan Bradley, yet another undergraduate in genetics at Cambridge.

half-time working for the EMBL at the EBI in Hinxton. My involvement in the genomics of *Drosophila* continued, as has my work with the Gene Ontology consortium, which has now matured into a major project. **Takis Benos** left the EBI to go to St. Louis to do a second postdoc with Gary Stormo. He now has a position at the University of Pittsburgh. He claims that Pittsburgh's reputation is undeserved. I am prepared to believe him rather than pay a return visit. **Ewan Birney** joined the EBI as a group leader in 2000 and has lead the very successful ENSEMBL project, a portal to the analysis of the human and other genomes. **David Botstein** left Stanford to head the Lewis-Sigler Institute at Princeton, where he is said to be very happy. He remains a force of nature. **Sydney Brenner** won the Nobel Prize for his work on *C. elegans* in 2002 and now has a new high-throughput sequencing venture, Population Genetics Technologies Ltd., in Cambridge. **Ken Burtis:** To his friends' dismay, is now dean of biology at the University of California, Davis. **Heather Butler** left FlyBase in 2001 to work for Caprion, a Canadian biotech company specializing in proteomics. **Graham Cameron** is now deputy director of the EBI in Hinxton. **Sue Celniker** remains at the Lawrence Berkeley Laboratory and has been responsible for all of the "finishing" of the *Drosophila* genome sequence. **Mike Cherry** still runs the yeast database at Stanford, and remains, with Suzi Lewis, Judy Blake from the Jackson Lab, and I, a PI on the Gene Ontology project, which the NIH now funds. **Peter** and **Lucy Cherbas** remain at Indiana University where Peter is now a professor of biology. **David Coates** is now dean of life sciences at the University of Bradford, United Kingdom. **Francis Collins**, a human geneticist, took over as director of the National Human Genome Research Institute from Jim Watson and has remained in the post, acting with Michael Morgan—as one of the two chief architects of the Human Genome Project, but also overseeing the funding of many public genome projects. **Richard Durbin** is a computational biologist extraordinaire at the Sanger nere he is deputy director to Alan Bradley. **Adam** d was Gerry's program officer at the NIH. He is an ɪary grant administrator, one of the unsung heroes who do so much work for the community and who make the NHGRI such a good institute with which to work. **Wolfgang Fleischman** remains at the EBI. **Mark Fortini** and **Debbie Morrison** remain married and both now have their own laboratories at the National Cancer Institute, Frederick, Maryland. **Bill Gelbart** is still the PI on FlyBase and still at Harvard, despite recurrent threats to go somewhere more civilized. **Richard Gibbs,** surprisingly for an Aussie, is faithful to Houston. He and his group have since sequenced the genome of a second species of *Drosophila*, and many less interesting organisms such as the rat and honeybee. **David Glover** remains Arthur Balfour Professor of Genetics at Cambridge, working to uncover the secrets of the cell cycle in flies and other organisms. **Phil Green** is now professor of genome sciences at the University of Washington in Seattle. **Greg Helt** joined Affymetrix when that company bought Neomorphic and telecommutes from his ranch in northern California. **Michael Hunkapiller** left Applied Biosystems in August 2004 after 21 years with the company. He now pursues independent business interests in biotech. **Fotis Kafatos** retired as director of the EMBL in April 2005 and is now leading a research group on malaria and its vectors at Imperial College, London. This is a little too close to Cambridge for my liking, but at least he no longer phones me at 10 p.m. on a Sunday night demanding my presence at a "vital" meeting in Brussels at noon the next day (or if he did, I could now say "no"). **Tony Kerlavage** was a vice president at Celera, having been, with Mark Adams, working with Craig since his days at the NIH. He remains at Celera. **Eric Lander** is now director of the Broad Institute in Cambridge, Massachusetts, his divorce from the Whitehead Institute at MIT being, I am sure, a relief to all concerned. **Bruno Lemaitre** now runs his own research group in the CNRS at Gif-sur-Yvette,

France. **Suzi Lewis** now runs her own computational biology group at the Lawrence Berkeley Laboratory, being free of the demands of FlyBase. **Peter Li** remains at Celera in Maryland. **David Lipman** remains head of the NCBI in Bethesda. **Rob Martienssen**, despite my best efforts to get him back home, stubbornly remains at Cold Spring Harbor Laboratory wresting secrets from plant genomes. **Dick McCombie** is still head of the sequencing lab at Cold Spring Harbor. **George Miklos** now lives on the beach in Australia, and is still—I assume –driving a Mercedes, and enjoying the good life. **Sima Misra** has remained with Suzi in Berkeley. In 2001, she led a complete reannotation of the genome of *Drosophila*, which was published in 2002. **Michael Morgan** retired from the Wellcome Trust in 2002 and, in recognition of his key role in establishing the Wellcome Trust Genome Campus at Hinxton, has been commemorated by the Morgan Building, which opened in October 2005. **Chris Mungall** still works in Suzi's group in Berkeley and has played major roles in the development of new databases for genomics. **Gene Myers** left Celera in September 2002 to rejoin academia as a professor in the computer science department at Berkeley. In late 2005, he joined Gerry at Janelia Farm. **Jim Ostell** remains at the NCBI in Bethesda. **Martin Reese** is now running a new start-up, Omicia, Inc., a genetics diagnostic company. Let's hope he does as well with this as he did with Neo. **Ken Rendle** has built Kenway Cars into a serious business. Kenway still does the run between Cambridge and Heathrow for me too often, although Ken himself has stopped driving. **Gerry Rubin**, as told, became vice president for biomedical research at the Howard Hughes Medical Institute (HHMI) in 2000 and has since become director of HHMI's Janelia Farm Research Campus, a magnificent new development for biology on a green field site just north of Dulles Airport. **John Sulston** won the Nobel Prize for Medicine or Physiology with Sydney Brenner and Bob Horvitz in 2002. Having lead the group at Hinxton that sequenced about one third of the human genome, he retired from the Sanger Centre (by then the Sanger Institute), wrote *The Common Thread*, and now does good work. **Craig Venter** was at the NIH where he pioneered the controversial EST sequencing program. Frustrated with lack of support, he left to found The Institute of Genomic Research (TIGR). Founded Celera in 1998. In January 2002, he was fired by Tony White and set up another not-for-profit institute, the J. Craig Venter Institute, in Rockville, Maryland. He and **Claire Fraser** divorced in the summer of 2003. Claire took over as director of TIGR when Craig founded Celera, and she remains there, growing TIGR into a major sequencing and genomics center. Craig has made great use of his yacht *Sorcerer II* to sail around the world collecting, and sequencing, microbial samples from the oceans. **Jim Watson** is chancellor of Cold Spring Harbor Laboratory, raising money for an extensive new development. **Jennifer Wortman** left Celera in 2002 and is now bioinformatics manager of eukaryotic annotation at TIGR. **Mark Yandell** left Celera in February 2002 to join Gerry Rubin's group in Berkeley. In August 2005, he and his wife Karen Eilbeck, also from Celera, moved to their own positions at the University of Utah.

Epilogue

Science needs heroes and heroic stories. Michael is not a hero, but his story is heroic. Very few scientific stories emerge naked from a bath, and this is no exception. From day to day our understanding of the world is built incrementally, brick by brick on what are, we trust, firm foundations. Every now and again a wall, or even a whole room, falls down, because its foundations were insecure, but such "paradigm shifts"[1] are rare; most of us have a fruitful scientific life without first-hand experience of such a revolution. But behind Michael's story are four revolutions in biology. The first, of course, was Darwin's theory of evolution by natural selection, published in 1858, a theory which, to paraphrase Theodosius Dobzhansky "makes sense of biology."[2] The second, in 1900, was the rediscovery of Mendel's work of 1866, in which he elucidated the mechanisms of heredity, allowing an understanding of the process of evolution.[3] The third, in 1953, was the discovery of the structure of DNA, by

---

[1] A shorthand used by Kuhn in his *The structure of scientific revolutions,* University of Chicago Press, 1962.

[2] See, for example, Dobzhansky T. 1973. Nothing in biology makes sense except in the light of evolution. *Am. Bio. Teacher* 35: 125–129.

[3] For an accessible history of genetics, see Sturtevant A.H. 1965. *A history of genetics.* Republished with an introduction and afterward by E.B. Lewis, Cold Spring Harbor Laboratory Press and Electronic Scholarly Publishing Project, 2001.

James Watson and Francis Crick, which gave a molecular basis to these mechanisms. The final revolution was the development of methods to clone and sequence DNA. Many contributed to this revolution, itself a direct result of the rise of molecular biology, following Watson and Crick's discovery of the structure of DNA.

Between revolutions, we do what Kuhn, somewhat disparagingly, called "normal science." This term hides the fact that although many discoveries in science are obvious (at least with the benefit of hindsight), many are not and result in local civil unrest, rather than global revolution. Discoveries come in many forms. One that has been very important in the history of genetics has been the "discovery" of the potential of particular organisms for experimental or observational study. Darwin's finches[4] and Mendel's peas are obvious examples. Both had, (and in the case of the former, continues to have) a major impact on our science. The "discovery" of *Drosophila melanogaster* by Thomas Hunt Morgan in about 1907 and, in particular, the discovery of a white-eyed mutation of this fly in May 1910, had an even greater impact. It was with this small, ubiquitous fly that Morgan, and his brilliant students, established the principles of classical genetics in the roughly period of roughly 1910–1936.[5]

Michael alludes to the competition between communities working with different organisms. This competition is for research funds and for the brightest students. It is also a competition for ideas and discoveries. Nobody (I hope) studies *Drosophila melanogaster* for its own sake; we study it because we hope to discover facts and principles that are universal. For instance, *Drosophila* biologists are proud that it was with *their* organism that Muller discovered (in 1928) that ionizing radia-

---

[4] See Grant P. 1999. *The ecology and evolution of Darwin's finches.* Princeton University Press, New Jersey.

[5] For a history of this period, see Kohler R.E. 1994. *Lords of the fly.* Chicago University Press.

tion induces genetic mutations; they are proud that certain classes of protein, of universal interest, were first discovered by biologists working with flies,[6] and they are proud that most of the major signal transduction pathways, so important for our understanding of human disease, were first discovered as a result of work with flies. Such discoveries bring more cash for their research, and the fact that these discoveries are widely discussed in the scientific press and at meetings inspires more students to join in research with *Drosophila*. But subtle forces are at play here. Science is an intensely competitive business. With major discoveries go self-esteem, promotion, and (perhaps) fame. Major discoveries in a lab attract the brightest students, who will, we hope, make major discoveries in their turn. This can result in great tension. Science is a very public activity. It is an activity that depends on the ability of others to repeat our research and build on it. Science can only progress if knowledge, and the material resources required for acquiring knowledge, are freely distributable to our colleagues—be they collaborators or competitors.

Resolving the tension between competition and the free exchange of information and materials has never been easy, as Michael's story makes very clear. There is, however, a historical background to this story, that Michael omits. It is that the founder of the field of *Drosophila* research, Thomas Hunt Morgan, was a very wise man. He realized that if research with *Drosophila* was to have a major impact on science, as many biologists as possible must be drawn into the field. Was this just an altruistic judgment, or did he realize that his glory would be enhanced if *Drosophila* research flourished?[7] In any case, the facts of the matter are clear: Morgan and his students con-

---

[6] For example, homeobox proteins and heat-shock proteins are but two of a long list.

[7] T.H. Morgan was awarded the Nobel Prize for Medicine or Physiology in 1933. For a biography of this remarkable man, see Allen G.E. 1978. *Thomas Hunt Morgan: The man and his science*. Princeton University Press, New Jersey.

sciously established what Kohler has called the *"Drosophila Exchange Network"*: They freely distributed their knowledge (very often before its publication) and their materials, i.e., their stocks of *Drosophila*, to any interested scientist and encouraged the use of *Drosophila* in teaching. This "open source" policy predates by decades the more modern debate about the ownership of intellectual "property" in science. There is no doubt in any of our minds that this policy was a major force driving research with *Drosophila*. It is a policy that we must fight, in today's world of institutional bureaucracy and "material transfer agreements," to maintain.

But what of the science? This is not the place for a history of *Drosophila* genetics. It is, however, the place to summarize the discoveries that lead to the determination and publication of the complete sequence of the genome of *D. melanogaster*. There are remarkably few key players here. Yes, many contributed, but few survive what Gerry Rubin has called the "deletion test": Would our science have progressed more slowly, or differently, had they never lived or at least never done science (the difference being minimal in this case).

The mutation responsible for the white-eye phenotype, discovered by Morgan in May 1910, showed a pattern of inheritance unexpected from Mendel's laws. It was *sex linked*, passed from a male only to his daughters.[8] This pattern of inheritance precisely paralleled that of the X chromosome, indeed W.S. Sutton had, in 1902, suggested that the chromosomes are the carriers of the hereditary material.[9] But the *proof* of the chromosome theory of heredity came from a remarkable student of

---

[8] Sex-linked inheritance was first discovered in a moth, *Abraxas*, in 1906: Doncaster L. and Raynor G.H. 1906. Breeding experiments with Lepidoptera. *Proc. Zool. Soc. Lond.* 1: 125–133.

[9] Sutton W.S. 1902. On the morphology of the chromosome group in *Brachystola magna. Biol. Bull.* 4: 24–29. Extraordinarily, Sutton worked not at Columbia or Johns Hopkins or Harvard, but in Kansas!

Morgan's, Calvin Bridges.[10] A sign often seen above the lab benches of fly pushers, attributed to William Bateson, the Cambridge man who brought Mendel's work to the English-speaking world, advises "Treasure your exceptions." How Bridges must have loved this, for it was the rare, exceptional behavior of the inheritance of the white-eyed character, in which (for example), a son inherited his father's X chromosome, that allowed Bridges to prove beyond doubt the chromosome theory of heredity. His thesis was published as the first paper in a new journal, *Genetics*.[11] Bridges had discovered the phenomenon of *nondisjunction*, resulting from errors in the meiotic segregation of homologous chromosomes. These are two wonderful examples of discoveries of universal impact: That genes are carried by chromosomes is a universal fact of life and that the meiotic segregation of chromosomes can go wrong is also a universal fact of life (for eukaryotes). That nondisjunction can have dramatic, and tragic, consequences in humans was a discovery that would have to wait until the late 1950s.[12]

Of course, no one in Morgan's group doubted the chromosome theory of heredity. Indeed, a discovery as fundamental as that of Bridges was made by a second brilliant student of Morgan's, Alfred H. Sturtevant. In Mendel's experiments, different genetic factors (for example, seed shape and color) were inherited independently. That this was not always true was discovered by Bateson and Punnett in 1906, but they worked with chickens. Sturtevant realized that the pattern and degree of nonindependence (linkage) between different genetic factors on the X chromosome of *D. melanogaster* could be understood

[10] See Lewis E.B. 2003. C.B. Bridges' repeat hypothesis and the nature of the gene. *Genetics* **164**: 427–431. A biography of the remarkable and colorful life of Calvin Bridges is long overdue.

[11] Bridges C.B. 1916. Nondisjunction as proof of the chromosome theory of heredity. *Genetics* **1**: 1–52.

[12] See Hawley R.S. and Mori C.A. 1999. *The human genome. A user's guide.* Academic Press, San Diego, California.

to ask to write the chapter on polytene chromosomes. This he
did, superbly. He accompanied them by a set of reference photographs of the polytene chromosome maps. George also had
the bright idea of ordering, at his own expense, 1000 reprints of
these from Academic Press, and, while stocks lasted, he and
Michael distributed them to anyone who asked (perhaps they
asked for a nominal charge of $1, but that could not have cov-
ered the costs). They made the interpretation of what was seen
in the microscope much more accessible to researchers than
had Bridges' drawings, fine though they are.

Whereas most of us were selfishly content to map our gene
of interest, other investigators were interested in understanding
the architecture of large chromosomal regions. George Lefevre
deserves great credit for being one of the first *Drosophila* geneticists to try to map an entire region of a chromosome rather
than just positioning a given mutant.[19] However, it was really
the work of Burke Judd and Michael Ashburner that would push
to its limits what was called the "saturation genetics analysis" of
large chromosomal regions, using a combination of genetic and
cytological analysis, to very high resolution.[20]

As these comprehensive surveys of relatively large chromosomal regions were published, many, including Michael,
became embroiled in a rather curious controversy surrounding
the idea that each polytene chromosome band may correspond to one gene.[21] This, the "one gene–one band" hypothesis, had a certain intellectual appeal and enormous conse-

---

[19] Lefevre G., Jr. 1971. Salivary chromosome bands and the frequency of
crossing over in *Drosophila melanogaster*. *Genetics* **67**: 497–513.

[20] Judd B.H., Shen M.W., and Kaufman T.C. 1972. The anatomy and function of a segment of the X chromosome of *Drosophila melanogaster*. *Genetics*
71: 139–156; Woodruff R.C. and Ashburner M. 1979. The genetics of a
small autosomal region of *Drosophila melanogaster* containing the structural
gene for alcohol dehydrogenase. II. Lethal mutations in the region. *Genetics*
92: 133–149.

[21] For an excellent discussion of the origin of this idea, see Judd B.H. 1998.
Genes and chromomeres: A puzzle in three dimensions. *Genetics* **150**: 1–9.

quences. Conventional genetic analysis had lead many to believe that there were only 5,000 or so genes in flies; the same number as that of chromosome bands. Moreover, the saturation studies of small regions, such as that of Burke Judd and colleagues in the *zeste-white* region of the X chromosome and Michael and colleagues in the *Adh* region of the second chromosome, also found a 1:1 ratio of genes and bands. True, there were some exceptions, but why throw away a beautiful hypothesis just because of the data? The flaw is (now) obvious: The great majority of the genes that were studied in these chromosome regions were vital genes, genes that when mutant were lethal to the fly. At the time, we had no idea what proportion of the genes in the fly were vital; common sense lead us to believe that it may be "most." But we were wrong. As Michael and Gerry showed in their 1999 paper on the sequence of the *Adh* region,[22] in fact only 25–30% of the genes in *D. melanogaster* are absolutely required to be intact for an adult fly to live. The numerical coincidence between the number of genes and the number of vital genes is simply that, a coincidence.

I have said that the existence of polytene chromosomes allowed fly geneticists to map genes with, as we know today, a resolution of 100 kb or so—a fact that lead to considerable "fly envy" among those investigating other organisms. But that resolution could only be achieved under very good circumstances, in particular, having a very large array of partially overlapping deletions and superb skill at "reading" the polytene chromosomes.

Obviously, such speculations about the relationship of genes and bands cried out for a way to map not just genes but

---

[22] Ashburner M., Misra S., Roote J., Lewis S.E., Blazej R., Davis T., Doyle C., Galle R., George R., et al. 1999. An exploration of the sequence of a 2.9-Mb region of the genome of *Drosophila melanogaster*: The *Adh* region. *Genetics* 153: 179–219; see also Lefevre G. and Watkins W. 1986. The question of the total gene number in *Drosophila melanogaster*. *Genetics* 113: 869–895.

also DNA sequences onto polytene chromosomes. Here, two major developments proceeded in parallel and greatly synergized each other. The more fundamental was the application of the newly discovered methods to clone DNA sequences to *Drosophila*. In the early 1970s, this was driven by David Hogness who had the good fortune to be at the epicenter of this technological development, the Stanford Medical School. Dave had been a distinguished phage molecular biologist who, in 1968, took a year off to learn about *Drosophila*. He spent time in three great labs: Ed Lewis's in Caltech, Jim Peacock's in Canberra, and Wolfgang Beermann's in Tübingen, and in 1969 he set up a fly lab in Stanford. Here, he attracted a brilliant group of graduate students and postdoctoral fellows, not the least of which was Gerry Rubin. It was within this group that we can say that genomics, as a science, really began. Not only were the first genome libraries constructed and the first *Drosophila* genes cloned, but so were the first eukaryotic transposable elements identified and the first attempts to use genomics to understand complex genetic problems, for example, the structure of the *Bithorax* complex, so beloved of Ed Lewis, made. Many major technical advances were made, including that of "chromosome walking," using the overlap between adjacent cloned sequences to cover large chromosome regions.[23]

One of the wonderful gifts of nature is the sequence complementarity of double-stranded DNA. This means that if double-stranded DNA is experimentally uncoiled (a very simple procedure in the lab), the single strands will hybridize with any strand of complementary sequence. If this complementary sequence is labeled in some way, for example, with a radioac-

---

[23] Bender W., Spierer P., and Hogness D.S. 1983. Chromosomal walking and jumping to isolate DNA from the *Ace* and *rosy* loci and the *Bithorax* complex in *Drosophila melanogaster. J. Mol. Biol.* **168:** 17–33.

tive isotope, one can locate sequences that are arrayed on a substrate. Mary Lou Pardue, then working with Joe Gall in Yale, realized that chromosomes could be just such a substrate. If the double-stranded DNA of chromosomes could be uncoiled (without destroying chromosome structure), "probing" it with a radioactive DNA molecule would allow that molecule to be mapped to the chromosomes. The method of in situ hybridization was born.[24] The combination of the cloning technologies pioneered by Dave Hogness and his group, and the ability to map cloned sequences with single polytene chromosome band resolution, enabled by in situ hybridization, revolutionized the practice of *Drosophila* biology. But more was to come.

Griffith had discovered DNA transformation in bacteria in 1928—but he did not know it. Indeed, it was the evidence that his "transforming principle" was DNA, a discovery of Avery, MacLeod, and McCarty in 1944, that began to convince the world that DNA, and not protein, was the genetic material.[25] By the 1970s, DNA transformation—the introduction of DNA molecules into an organism and their integration into their host's DNA—was a routine tool for the genetic analysis of bacteria.[26] Drosophilists do not have *coli* envy; *coli* are, after all, merely bacteria. But they *do* have yeast envy—yeast are eukaryotes. The successful genetic transformation of yeast, by Jean

---

[24] Pardue M.L., Gerbi S.A., Eckhardt R.A., and Gall J.G. 1970. Cytological localization of DNA complementary to ribosomal RNA in polytene chromosomes of Diptera. *Chromosoma* 29: 268–290.

[25] Avery O.T., MacLeod C.M., and McCarty M. 1944. Studies on the chemical transformation of pneumococcal types. *J. Exp. Med.* 79: 137–158. See also McCarty M. 1985. *The transforming principle. Discovering that genes are made of DNA.* W.W. Norton, New York.

[26] The ability to transform *E. coli* with exogenous DNA was critical for the development of recombinant DNA technology.

Beggs and Gerry Fink's group in 1979,[27] caused envy and consternation. Would the best students now go and work in yeast labs? The challenge of achieving DNA tranformation in flies had to be met. Many tried. Many failed. Many cried. Michael tried. Michael failed. Perhaps he cried. In 1982, fly geneticists met at the University of Connecticut in Storrs. Rarely do standing ovations occur at scientific meetings. But this time, Gerry Rubin and Allan Spradling got one, for they had successfully transformed *Drosophila*, and—true to Morgan's tradition—offered their methods and flies to all comers.

The success of Gerry and Allan had built on the discovery of a weird and obscure phenomenon called *hybrid dysgenesis*. Two groups, Margaret Kidwell's in the United States and Picard and L'Heritier in France, had independently discovered that when certain strains of flies were crossed together, their progeny were sterile.[28] Strains were of two types, called P and M by Margaret and I and R by the French team and the basic rule was simply that dysgenesis was only seen in the progeny of a cross of P (or I) males crossed to M (or R) females. Dysgenesis was never seen in any other cross between strains. For some time, there was total confusion because the phenomenology of the American and French dysgenesis differed greatly. Soon, in a display of common sense unusual today between these countries, the American and French groups got together and showed that PM and IR dysgenesis are independent systems. It was the PM system that was to have such a major role in our lives. Mel Green, a grand old man in the grandest tradition of fly genetics, had been suggesting for some time that mobile genetic elements might be responsible for some mutations in

---

[27] Beggs J.D. 1978. Transformation of yeast by a replicating hybrid plasmid. *Nature* **275**: 104–109; Hinnen A., Hicks J.B., and Fink G.R. 1978. Transformation of yeast. *Proc. Natl. Acad. Sci.* **75**: 1929–1933.

[28] For a review, see Chapter 30 in Ashburner M., Golic K.G., and Hawley R.S. 2005. Drosophila: *A laboratory handbook*, 2nd edition. Cold Spring Harbor Laboratory Press, New York.

flies. Gerry and Margaret collaborated to determine if this was so for the mutations commonly seen as a consequence of PM dysgenesis. They exploited the fact that the *white* gene, of T.H. Morgan fame, had recently been cloned by Bob Levis and Paul Bingham in Gerry's lab.[29] They isolated an allele of *white* induced by PM dysgenesis and showed that it carried an insertion of a novel DNA sequence, a DNA sequence present in all P strains but absent from any M strain. They named it the *P-element*.[30] The *P-element* has proven to be an unbelievably powerful tool for *Drosophila* geneticists. It can be used as a mutagen, and—a fact of great importance—because it marks (by its insertion) the gene it mutates, it allows the trivial molecular cloning of that gene. The *P-element* is a transposable element in the genome of the fly; that is why it can cause hybrid dysgenesis. It moves because it encodes an enzyme, a transposase, that catalyses its ability to cut itself out from one region of the genome and paste itself into another. It is this property, that allowed Gerry and Allen to use it as the basis of a vector for genetic transformation.[31] From now on, we had the ability to insert genes, perhaps after modification, into the germ-line chromosomes of *Drosophila* at will. No longer were we envious of yeast.

Used in various combinations, these methods were really quite efficient and, as noted by Rubin and Lewis (see footnote 17), by the end of the 20th century, more than 1300 *Drosophila* genes, the vast majority of which were originally identified by genetic analysis, had been cloned and sequenced by individual

---

[29] Bingham P.M., Levis R., and Rubin G.M. 1981. Cloning of DNA sequences from the *white* locus of *Drosophila melanogaster* by a novel and general method. *Cell* 25: 693–704.

[30] Rubin G.M., Kidwell M.G., and Bingham P.M. 1982. The molecular basis of P-M hybrid dysgenesis: The nature of induced mutations. *Cell* 29: 987–994.

[31] Rubin G.M. and Spradling A.C. 1982. Genetic transformation of *Drosophila* with transposable element vectors. *Science* 218: 348–353.

labs. For most of us, the goal of such exercises was simply to find our gene of interest by whichever combination of tools seemed most expedient. Other genes in the region were simply suspects to be ruled out or walked through. But a few labs did the huge amount of work that tied cloning and DNA sequence analysis to the sort of saturation analysis of large polytene regions that I describe above. Most notable among these efforts was the molecular analysis of region including the *Adh* gene that was performed by Michael Ashburner and colleagues. In 1985, Ashburner and coworkers published an analysis of the 165-kb interval containing *Adh* as well as two other genes, *outspread* and *no-ocelli*.[32] This study provided one of the first real glimpses into the complexity of gene organization in *Drosophila* by demonstrating that the *outspread* gene spanned at least some 52 kb in length (indeed, we now know that it is closer to 90 kb!) and fully encompasses the *Adh* gene. As a presage to sequencing the genome, Ashburner, Rubin, and their colleagues then extended this combination of detailed genetic and molecular characterization to dissect a 2.9-mb region, also spanning the *Adh* gene.[33] This region spanned about 70 polytene bands and the resulting sequence analysis suggested that it contained 218 protein-encoding genes, and 144 (66%) of these genes encode proteins homologous to proteins in other organisms. The region was punctuated by 650 aberration breakpoints and, of 73 genes identified in the region by genetic analysis, 43 were localized on the sequence. Studies such as this made it clear that genomic analysis was clearly possible in *Drosophila*, awaiting only a completed sequence.

Unfortunately, the success of gene cloning as a cottage industry in the fly community somewhat lessened the commu-

---

[32] Chia W., Karp R., McGill S., and Ashburner M. 1985. Molecular analysis of the *Adh* region of the genome of *Drosophila melanogaster. J. Mol. Biol.* **186:** 689–706.

[33] See footnote 22.

nity's desire to have a complete genome sequence. As written by Rubin and Lewis (see footnote 17), "Ironically, the success in cloning and studying individual genes dampened enthusiasm for an organized genome project, which was seen as unnecessary." But those who failed to see the necessity of having that sequence were missing one of the major points of Ashburner and colleague's paper, namely, that of the 218 protein-coding genes in this 2.9-mb interval, only 73 (approximately one third) of these genes were defined by the saturation genetic analysis performed by Ashburner and his colleagues in their studies over a period of 20 years or so. Thus, the majority of genes in the fly genome might not be identifiable by the process of cloning genes defined by mutations with defects of interest. Moreover, given the progress of sequencing programs in other organisms, the use of *Drosophila* as a valuable model system clearly depended on our ability to rapidly identify and characterize the fly homologs of genes of interest in other organisms. By this time, fly workers were beginning to suffer a serious case of not only yeast envy, but also of worm envy.[34] Clearly the sequence of *D. melanogaster* was essential if we were to be able to hold our heads high, and in collaboration with Craig Venter and a company called Celera, Gerry Rubin and his collaborators delivered that sequence in 2000. That's the story told in this book.

But I do not want to finish without providing a clear statement of why the sequence matters. Before the release of the sequence and the tools it has enabled, I used to tell those in my lab that cloning a gene defined only by an ethylmethanesulfonate-induced mutant was one postdoctoral lifetime's worth of work. But having the sequence available has created techniques such as single-nucleotide polymorphism mapping and a

---

[34] The complete sequence of the yeast *Saccharomyces cerevisiae* was completed by May 1997 (see *Nature,* May 29, 1997, Supplement) and that of the worm *Caenorhabditis elegans*, by December 1998 (see *Science*, December 11, 1998).

collection of small deletions with precisely defined break-points. Earlier this year, my lab completed a screen for meiotic mutations that yielded 11 novel mutants. As of the date of this writing (December, 2005) two thirds of these mutations have been mapped to a gene and the molecular lesion identified. Without the sequence, that would not have been possible. On the other side of the coin, most genes identified by sequencing alone become interesting only when genetic or cell biological tools provide a means to identify their function. That is why this story is so important.

R. SCOTT HAWLEY

Stowers Institute for Medical Research
Kansas City, Missouri

Afterword

I have always admired Michael and his steadfast devotion to flies, but I must say that this narrative reveals a very different facet of him, one that is a hard act to follow. Writing a sensible epilogue to this tale is a bit like Ed McMahon playing straight man to Johnny Carson. Be that as it may, I believe that it is fair to state from the outset that Michael and his genome compatriots have made a great contribution to the *Drosophila* field. In contrast to some of the large egos that dominated the press, admittedly with good justification, Michael's passion for *Drosophila* has left a mark that might have gone unappreciated by the many fly people who otherwise could have simply accepted the powerful well-crafted genomic tools, such as detailed gene annotations and the associated ontologies, as givens had this inner story not been told in all its *gory*. So, I would like to begin by personally thanking Michael and others who selflessly added flesh to the *Drosophila* genome bones behind the scenes.

There have been several important immediate consequences of having the complete genomic sequences of *Drosophila* and other organisms in the past five years. There is also much more that is likely to come in the next decade. I will first summarize some of the most important contemporary applications of genome-scale information and then consider some possible future developments.

MATCHING GENES WITH FUNCTIONS

Drosophila was a leading organism for establishing gene function long before functional genomics became trendy. As Scott Hawley writes, completion of the genome sequence has greatly accelerated the identification of mutated genes in *Drosophila*, and it should soon be possible to obtain an allelic series of mutations in most genes. When researchers isolate a new mutation affecting a process of interest, they can rapidly map it to a chromosomal region encompassing ten or fewer genes by one of several effective methods such as the recombination system developed by Hugo Bellen using a defined set of well-spaced dominantly marked transposons, male recombination as pioneered by Bill Engels using single nucleotide polymorphisms (SNPs), or by complementation analysis with a collection of deletion stocks spanning the great majority of the genome.[1] It is also possible to use recombination methods to map the mutation to yet a finer scale, often to a single transcription unit. Candidate genes can then be sequenced to identify the mutation. All of these techniques are greatly facilitated by having the complete genome sequence in hand.

One of the most pressing objectives of *Drosophila* biologists, which is now more than half complete, is to create a collection of mutations in all predicted genes. The most widely used method for obtaining a mutation in a gene is to search for a known transposon insertion into or next to that gene, using either *P-elements* (see the Epilogue) or transposons such as *piggyBac*. Having the genome sequence has been essential for this project, because short sequences at the site of transposon insertion can be sequenced and matched to the assembled genome sequence to determine the location of that insertion. Allan Spradling, Hugo Bellen, and Gerry Rubin have put

---

[1] For a recent review of some of these technologies, see Venken K.J. and Bellen H.J. 2005. Emerging technologies for gene manipulation in *Drosophila melanogaster*. *Nat. Rev. Genet.* **6**: 167–178.

together a consortium that has been working hard to generate a nearly complete collection of such insertional mutations.[2] These screens should soon be approaching saturation. Even in cases for which the transposon insertion does not disrupt gene function, it is usually straightforward to screen for imprecise excisions of the marked transposon to obtain loss-of-function alleles of the gene. We can now foresee the day when the great majority of fly genes have at least one mutant allele—an essential prerequisite for determining their function.

The existing collection of deletion stocks covers about 85% of the genome. The *Drosophila* Deletion Project (DrosDel) and others[3] have been creating a collection of stocks that will permit users to generate designer deletions between defined chromosomal end points mapped to base-pair resolution, which will be a great boon for mapping mutations. Kits of these deletion stocks can also be used in second-site modifier screens to identify new candidate loci involved in a wide variety of biological processes.

Another powerful recent method for obtaining mutations in a gene of interest, including only partial loss-of-function alleles, is known as Tilling.[4] This method, which was first implemented

[2] See Bellen H.J., Levis R.W., Liao G., He Y., Carlson J.W., Tsang G., Evans-Holm M., Hiesinger P.R., Schulze K.L., Rubin G.M., Hoskins R.A., and Spradling A.C. 2004. The BDGP gene disruption project: Single transposon insertions associated with 40% of *Drosophila* genes. *Genetics* **167**: 761–781.

[3] Ryder E., Blows F., Ashburner M., Bautisa-Llacer R., Couslon D., Drummond J., et al. 2004. The DrosDel collection: A set of *P-element* insertions for generating custom chromosomal aberrations in *Drosophila melanogaster*. *Genetics* **167**: 797–813; Parks A.L., Cook K.R., Belvin M., Dompe N.A., Fawcett R., Huppert K., et al. 2004. Systematic generation of high-resolution deletion coverage of the *Drosophila melanogaster* genome. *Nature Genet.* **36**: 288–292.

[4] Colbert T., Till B.J., Tompa R., Reynolds S., Steine M.N., Yeung A.T., et al. 2001. High-throughput screening for induced point mutations. *Plant Physiol.* **126**: 480–484.

in plants, has now been adapted to *Drosophila*[5] and promises to deliver an average of ten mutant alleles of gene. At the heart of this effort is a collection of over 10,000 balanced stocks created in Charles Zuker's lab carrying chromosomes that were heavily mutagenized with EMS[6] (or it can be done with any other set of mutagenized chromosomes). Tilling systematically scans for point mutations in defined regions of the chromosome in DNA samples from each of the Zuker lines by identifying short fragments from mutant chromosomes with single base-pair mismatches to a wild-type reference genome. In principle, this screening could be performed on all genes to create a genome-wide repository of stocks carrying point mutant alleles that could be made available from a stock center.

The genome sequence has also played a crucial part in the development of genome-wide RNA interference (RNAi) screens, in which cellular phenotypes have been scored for the reduction/elimination of function for every predicted gene. This whole-genome RNAi technology, pioneered by Rich Carthew and Jack Dixon,[7] and then developed as a tool for genome-wide screening in cell culture by Norbert Perrimon and colleagues,[8] has lead to the identification of whole cadres of new genes involved in regulating the cytoskeleton, cell migration, and the

---

[5] http://tilling.fhcrc.org:9366/fly; Winkler S., Schwabedissen A., Backasch D., Bokel C., Seidel C., Bonisch S., et al. 2005. Target-selected mutant screen by TILLING in *Drosophila. Genome Res.* **15**: 718–723.

[6] Koundakjian E.J., Cowan D.M., Hardy R.W., and Becker A.H. 2004. The Zuker collection: A resource for the analysis of autosomal gene function in *Drosophila melanogaster. Genetics* **167**: 203–206.

[7] Kennerdell J.R. and Carthew R.W. 1998. Use of dsRNA-mediated genetic interference to demonstrate that *frizzled* and *frizzled 2* act in the *wingless* pathway. *Cell* **95**: 1017–1026; Clemens J.C., Worby C.A., Simonson-Leff N., Muda M., Machama T., Hemmings B.A., and Dixon J.E. 2000. Use of double-stranded RNA interference in *Drosophila* cell lines to dissect signal transduction pathways. *Proc. Natl. Acad. Sci.* **97**: 6499–6503.

[8] For example, see Friedman A. and Perrimon N. 2004. Genome-wide high-throughput screens in functional genomics. *Curr. Opin. Genet. Dev.* **14**: 470–476.

response to various signaling pathways. The cellular RNAi phenotype of a gene can be investigated further by classical loss-of-function studies in the whole fly to determine the organismic role of that gene. This is but one example of the fact that the availability of the entire sequence of the fly's genome has enabled very high-throughput technologies to be employed to dissect gene function. Others are discussed below.

The flip side of loss-of-function genetics is to overexpress a gene in its wild-type pattern or to misexpress it in cells that do not normally express that gene. Pioneering genomic screens performed by a collaboration between Dan Lindsley and Larry Sandler's groups in the early 1970s provided nearly genome-wide coverage of duplicated and deleted chromosomal regions in *Drosophila*.[9] Among the important insights derived from these seminal studies was the observation that flies could tolerate an extra copy of nearly every gene in the genome. A more recent method for supplying defined duplicated regions of the genome is known as recombineering.[10] The Bellen group is adapting this method to *Drosophila* from technology developed in the mouse that allows one to introduce 30–100-kb regions obtained from bacterial artificial chromosomes (BACs) into defined chromosomal sites. An important advantage of this system is that one can use various site-directed methods to mutate a BAC in bacteria (i.e., introduce point mutations or deletions between arbitrary end points) and then recombine these mutated sequences into the wild-type parent BAC, insert the altered BAC into the same chromosomal site in *Drosophila*, and compare the activities of the mutated and wild-type sequences. There are many possible uses of this method. For example, an investigator could map a mutation of interest down to a region encompassed by a BAC and show that it is possible to rescue that

---

[9] Lindsley D.L., Sandler L., Baker B.S., Carpenter A.T., Denell R.E., Hall J.C., et al. 1972. Segmental aneuploidy and the genetic gross structure of the *Drosophila* genome. *Genetics* **71**: 157–184.

[10] See footnote 1.

tion extracted as controlled vocabulary from abstracts or other text, the potential for extracting new information on networks of gene and protein interaction will gain even greater impact.

A related benefit of developing a controlled vocabulary for describing *Drosophila* phenotypes is that one can then compare information obtained from different species such as *C. elegans*, yeast, mice, or humans to identify potentially unanticipated intersections of pathways. One cross-genomic database, Homophila,[16] allows users to search for *Drosophila* homologs of human disease genes listed in the Online Mendelian Inheritance in Man[17] (OMIM) by keyword or gene name. Such cross-genomic analysis can be viewed as translations from one genetic language to another (e.g., *Drosophila* to human genetics). For example, lexical items in the language of *Drosophila* genetics used to describe *Notch* pathway function include terms such as wing margin, vein thickness, and embryonic cuticle holes. These terms are restricted to *Drosophila* because humans do not have wings and nor a cuticle covering their epidermis. On the other hand, in humans the lexicon for *Notch* signaling includes terms such as vertebral anomalies, spondylothoracic dysplasia and dystosis, and arteriopathy. These terms clearly are not relevant to *Drosophila* which does not have a bony endoskeleton or closed vasculature. Nonetheless, one could ask whether any *Drosophila* gene associated with a fly-specific *Notch*-related phenotype might be involved in an inherited human skeletal malformation and if such a homolog were mapped within a candidate region for the gene in question, it would certainly rise to the top of the list of candidate genes for causing that disease. As efforts at creating more detailed controlled vocabularies move forward, such cross-genomic translations will become increasingly effective means for combining information about genetic pathways and their network interactions with other pathways.

---

[16] http://superfly.ucsd.edu/homophila
[17] www.ncbi.nlm.nih.gov/entrez/query.fcgi?db=OMIM

Another potentially powerful use of cross-genomic comparisons is to identify genes carrying unique organism-specific functions. For example, Charles Zuker and colleagues made use of the fact that ancestral eukaryotes were ciliated and that cilia were lost independently in several different lineages to define a set of genes present only in genomes from ciliated organisms.[18] This lead to the identification of several genes involved in the transport of proteins into cilia, as well as an explanation for the functions of several independent loci causing Bardet-Beidl syndrome.

TRACKING DOWN POLYGENIC TRAITS

One important developing area in *Drosophila* genetics is the application of quantitative trait-mapping schemes and congenic breeding strategies to identify loci contributing to phenotypes including morphology, physiology, and behavior. These are "difficult" phenotypes, because they are determined by variation at many genes and are characterized by strong genotype-environment interactions. Yet their study is critical if we are to understand the basis of many common human diseases, such as heart disease and diabetes. This growing area makes use of classical methods in population genetics and mapping molecular markers such as single-nucleotide polymorphisms. Cross-genomic comparisons are also likely to have an important role in analyzing multigenic traits. For example, as the human HapMap initiative moves forward in mapping human haplotypes and associating multiple loci with diseases,[19]

[18] Avidor-Reiss T., Maer A.M., Koundakjian E., Polyanovsky A., Keil T., Subramaniam S., and Zuker C.S. 2004. Decoding cilia function: Defining specialized genes required for compartmentalized cilia biogenesis. *Cell* **117:** 527–539.

[19] Altshuler D., Brooks L.D., Chakravarti A., Collins F.S., Daly M.J., Donelly P., et al. (The International Hapmap Consortium). 2005. A haplotype map of the human genome. *Nature* **437:** 1299–1320.

Drosophila is uniquely poised to test combinations of genotypes for the additive or synergistic contributions to a given phenotype. It is also possible in some cases to cross closely related drosophilids that exhibit distinct behaviors (e.g., courtship) or environmental characteristics (e.g., resistance to drought), which opens the way to mapping loci contributing to such complex phenotypes.

HETEROCHROMATIN AND EPIGENETIC PHENOMENA

One of the remaining genome sequence goals is to obtain a complete sequence for heterochromatic regions of the genome. This is a nontrivial effort since much of this sequence is highly repetitive, confounding assembly efforts. There are important reasons to push forward to forge past these challenges, however, because several known loci with important functions map to the heterochromatin. In addition, essential chromosomal structures, such as centromeres, map to heterochromatic regions. An important observation made by Gary Karpen and colleagues[20] regarding the nature of centromeres is that they appear to be defined by largely epigenetic phenomena in that DNA sequences near a functional centromere can acquire centromeric activity in a manner that does not appear to depend on primary sequence. This is a fascinating phenomenon that may also be relevant to chromosomal propagation in other organisms.

MICROARRAYS

One of the most powerful genome-scale technologies is the microarray. Pioneered by researchers at Stanford and Affy-

---

[20] See, for example, Sun X., Le H.D., Wahlstrom J.M., and Karpen G.H. 2003. Sequence analysis of a functional *Drosophila* centromere. *Genome Res.* 13: 182–194; see also http://www.dhgp.org/.

metrix, this method allows one to examine changes in the transcription level of all predicted genes that are on the order of two- to threefold or greater. Detailed comparisons between expression profiles collected during embryonic, larval, and pupal development have revealed interesting trends and facts. For example, the first such comprehensive analysis performed by Matt Scott and colleagues revealed that approximately 90% of all genes are expressed at some time during embryonic development and that separate subsets of genes are expressed during embryonic versus larval development.[21] By comparing embryos in which too few or too many cells adopted mesodermal fates, this group also identified many new genes expressed in the developing mesoderm and identified a new important regulator of mesodermal development.[22] Similar experiments performed by Mike Levine's lab identified a new set of genes expressed in the developing neuroectoderm. Two of these new genes turned out to encode the long-sought ligands for the fibroblast growth factor (FGF) receptor *Heartless*, required for mesoderm migration beneath the neuroectoderm and for formation of the visceral mesoderm and heart.[23] Similar comparative analyses have identified important cohorts of coregulated genes in other important biological processes such as circadian rhythm and the immune responses to bacterial and/or fungal pathogens.

Another informative use of microarrays is to define sets of genes that are coregulated in a variety of different settings.

[21] Arbeitman M.N., Furlong E.E., Imam F., Johnson E., Null B.H., Baker B.S., et al. 2002. Gene expression during the life cycle of *Drosophila melanogaster*. *Science* 297: 2270–2275.

[22] Furlong E.E., Andersen E.C., Null B., White K.P., and Scott M.P. 2001. Patterns of gene expression during *Drosophila* mesoderm development. *Science* 293: 1629–1633.

[23] Stathopoulos A., Tam B., Ronshaugen M., Frasch M., and Levine M. 2004. *pyramus* and *thisbe*: FGF genes that pattern the mesoderm of *Drosophila* embryos. *Genes Dev.* 18: 687–699.

Such genes may function as a unit to perform a common function. Because coregulated genes are likely to respond to common upstream regulators, they can be analyzed by computational methods (see below) to identify potential shared *cis*-regulatory inputs (e.g., common transcription factor binding sites). A systematic effort is now underway by Mark Biggin and colleagues at Berkeley to determine the DNA-binding specificities of all predicted transcription factors in *Drosophila*.[24] This type of analysis is made yet more powerful when *cis*-regulatory sequences from multiple *Drosophila* species are aligned. Genome sequences with moderate coverage also exist for 11 *Drosophila* species including species that are closely related to *D. melanogaster* as well as others more distantly related. As discussed further below, highly conserved noncoding regions often contain clusters of transcription factor binding sites that correspond to enhancer elements.

MASS-SPEC PROTEOMICS

Having personally tried in the past to identify a new protein in *Drosophila* using a partially purified preparation of immune precipitated protein, I am truly amazed by the power of modern proteomics. It is now routine to submit barely detectable quantities of a protein to analysis by tandem mass spectrometry and get an unequivocal identification of that protein by comparing its observed proteolytic fragmentation pattern and limited amino acid sequence with the predicted patterns and sequence of known proteins in the genome. It sounds like magic, but it really works. This method can also be used to identify proteins present in complex mixtures and is very effective in determining the constituents associated in a partially purified protein complex. The applications of this methodology are largely limited by imagination, but include identification of protein targets that interact with or are modified in some fashion (e.g., phos-

---

[24] http://bdtnp.lbl.gov/

phorylated, ubiquitinated, or cleaved) by a protein of interest, identification of the constituents of alternatively formed protein complexes, and the identification of proteins that differ in expression level or state in different cell types or in response to added extracellular signals.[25]

Systematic screening for protein–protein interactions in yeast is another important proteomic application of the genome sequence. For example, analyses of all possible pairwise interactions between predicted *Drosophila* proteins (as assayed in the nucleus) has helped many researchers identify new components of systems that they have been studying.[26] Similar types of screens have been developed to identify protein–protein interactions taking place in the membrane or extracellularly.[27] Another powerful proteomic method, known as chromatin immunoprecipitation (ChIP), which detects interactions between proteins and DNA in vivo, makes important use of genome sequence because small fragments of genomic DNA that are isolated by virtue of their interaction with a given protein in nuclei are identified by hybridization to genome-wide microarrays. This type of "chip-on-chip" experiment can reveal downstream target genes regulated by a given transcription factor and can be used to determine the composition of transcriptional or chromatin complexes forming at defined sites in the genome.[28] Although ChIP was developed by

---

[25] Aebersold R. and Mann M. 2003. Mass spectrometry-based proteomics. *Nature* 422: 198–207.

[26] Giot L., Bader J.S., Brouwer C., Chaudhuri A., Kuang B., Li Y., et al. 2003. A protein interaction map of *Drosophila melanogaster*. *Science* 302: 1727–1736; Formstecher E., Aresta S., Collura V., Hamburger A., Meil A., Trehin A., et al. 2005. Protein interaction mapping: A *Drosophila* case study. *Genome Res.* 15: 376–384.

[27] See, for example, Fields S. 2005. High-throughput two-hybrid analysis. The promise and the peril. *FEBS J.* 272: 5391–5399.

[28] See, for example, Blais A. and Dynlacht B.D. 2005. Constructing transcriptional regulatory networks. *Genes Dev.* 19: 1499–5111; Sikder D. and Kodadek T. 2005. Genomic studies of transcription factor-DNA interactions. *Curr. Opin. Chem. Biol.* 9: 38–45.

Rob White in the pregenomic era,[29] it is only now that the full power of this method can be seen.

DECIPHERING REGULATORY DNA CODES

One of the biggest challenges in making use of genomic sequence data is to identify *cis*-regulatory enhancer elements using computational methods. Several strategies have been developed to accomplish this goal, which when combined are showing promise. The first method is to identify clusters of known transcription factor binding sites by choosing factors that are likely to be important regulators (e.g., by looking for binding sites for gap gene proteins in *cis*-regulatory regions of pair-rule genes) and then setting a threshold for how many such sites should fall within a given interval of DNA.[30] The second approach, which makes use of DNA sequence information from several species of drosophilid, is to search for highly conserved islands of sequence conservation among closely related species or moderately conserved intervals among more distantly related species.[31] When these two independent methods for identifying potentially relevant sites are combined, a significant fraction of known enhancers are found and many de novo predicted regulatory elements turn out to function as real enhancer elements. Although these preliminary computational approaches are promising, additional advances will be required to increase the fidelity in identifying known regulatory elements and to reduce the number of false positives, which is still considerable.

[29] Gould A.P. Brookman J.J., Strutt D.I., and White R.A. 1990. Targets of homeotic gene control in *Drosophila*. *Nature* **348:** 308–312.

[30] Chan B.Y. and Kibler D. 2005. Using hexamers to predict *cis*-regulatory motifs in *Drosophila*. *BMC Bioinformatics* **6:** 262.

[31] http://pipeline.lbl.gov/cgi-bin/gateway2?bg=dm1; Frazer K.A., Pachter L., Poliakov A., Rubin E.M., and Dubchak I. 2004. VISTA: Computational tools for comparative genomics. *Nucleic Acids Res.* **32:** W273–W279.

CONSTRUCTING GENE EXPRESSION ATLASES

Microarrays offer the great advantage of being able to query the entire collection of expressed genes in a given experiment, however, they do not provide precise temporal or spatial information. On the other hand, although it is possible to achieve single-cell resolution over a timescale of 2–5 minutes of development using histochemical in situ RNA detection methods, these methods typically are used to examine the expression of only one or two genes at a time. More recently, a highly sensitive fluorescent-based in situ method has been developed by Dave Kosman and colleagues that allows one to examine the expression of up to seven different transcripts at a time in a single embryo.[32] This method can also be used to create a multiplex code for scoring nascent transcription foci (nuclear dots in the vernacular) that, in principle, should be able to detect as many as 20–50 transcripts per cell in a single embryo. Such multiplex methods help fill the gap between genome-wide microarrays and traditional histochemical-based in situ detection methods and can be used to determine multiple gene expression patterns in embryos of complex genotype. In the near future, these methods should enable the construction of genome-wide gene expression atlases, which will be essential in decoding regulatory relationships between networks of genes and pathways.

FUTURE FANTASIES

Although crystal balls are notoriously inaccurate tools for gazing into the future, extrapolating from existing methods and technologies one could hazard a guess as to what an online

---

[32] Kosman D., Mizutani C.M., Lemons D., Cox W.G., McGinnis W., and Bier E. 2004. Multiplex detection of RNA expression in *Drosophila* embryos. *Science* **305**: 846.

genome interface might look like a decade or two from now. If one imagines sitting at a computer running a rapid noncrashing operating system that also does not use any Microsoft program, what could be done to make the most of the wealth of information accruing on the function and interaction of genetic pathways? The data fields are already overwhelming and, as journals move toward open-access publication, how will your monitor be able to protect you from the crushing force of data flux? I certainly do not pretend to have any magic answers, but I think that the problem can largely be framed in terms of how to develop interfaces that allow one to explore data space from coherent logical perspectives. One could imagine clicking on a gene or pathway and seeing a representational globe appear that can be rotated around its various axes to reveal different continents of information such as controlled Gene Ontology profiles, phenotypes, expression profiles, genetic interactions, protein interactions, posttranslational states, regulatory network information, or cross-genomic relationships. By clicking on one of these general categories, or on defined country-like subdomains (e.g., whole-mount embryos under the category of expression), one could then be directed to various summary options (e.g., a page showing the expression pattern of a gene of interest in the embryo at different stages in combination with other genes that may also participate in a given process) and then by homing in on specific city-like features of that data spread (e.g., a quadruple stain of an embryo with genes involved in dorsal closure of the embryo) link back to the literature to find other points on the globe that have statistically nontrivial linkage to the data point of interest.

Two types of advances will be critical to creating such a global genome network system. First, the steady progress that is being made in the many current areas of research discussed above will need to approach saturation of information content (e.g., we will need to have complete collections of information regarding loss-of-function phenotypes for every gene in a wide

array of cell types and developmental stages and correspond-
ingly detailed combinatorial gene expression atlases). Second,
it will be necessary to invent new ways to organize and display
data for rapid user directed queries that will make it possible to
jet around the information globe in a free unimpeded fashion
to see the many sites and wonders of the genome world.
Although this may seem fanciful today, the future will most
likely surpass the imagination, as fusions of existing fields and
new eruptions of innovative thinking, typical of the *Drosophila*
field, redefine the tectonics of our ever-changing world. The
complete sequence of the genome of *D. melanogaster* is just the
beginning of an exciting journey of discovery. Back to the
bench!

ETHAN BIER

University of California, San Diego